Feminist Technosciences

Rebecca Herzig and Banu Subramaniam, Series Editors

Queer Feminist Science Studies

A READER

EDITED BY
CYD CIPOLLA
KRISTINA GUPTA
DAVID A. RUBIN
ANGELA WILLEY

UNIVERSITY OF WASHINGTON PRESS
Seattle and London

Printed and bound in the United States of America
Design by Thomas Eykemans
Composed in Chaparral, typeface designed by Carol Twombley
21 20 19 18 17 5 4 3 2 1

University of Washington Press
www.washington.edu/uwpress

Library of Congress Cataloging-in-Publication Data on file
ISBN (hardcover): 978-0-295-74257-1
ISBN (paperback): 978-0-295-74258-8
ISBN (ebook): 978-0-295-74259-5

The paper used in this publication is acid-free and meets the minimum requirements of American National Standard for Information Sciences—Permanence of Paper for Printed Library Materials, ANSI Z39.48–1984. ∞

Queering what counts as nature is my categorical imperative. Queering specific normalized categories is not for the easy frisson of transgression, but for the hope for livable worlds.

DONNA HARAWAY, "A GAME OF CAT'S CRADLE"

Contents

Acknowledgments

The editors would like to thank the various individuals, communities, and institutions that helped to make this project a reality. We are grateful to the College of Wake Forest University and Dean Susanne Wofford of the Gallatin School of Individualized Study for providing subvention funds. For their support and encouragement, we thank our colleagues and students in our home departments at University of Massachusetts Amherst, University of South Florida, Wake Forest University, and the Gallatin School of New York University. Additionally, we thank the audiences for our panels at the last several annual NWSA conferences for their insights and encouragement over the years.

We owe a profound debt of gratitude to Larin McLaughlin, editor in chief at the University of Washington Press, and Banu Subramaniam and Rebecca Herzig, the editors of the Feminist Technosciences series. Their faith in the project and guidance at every step of the process has been nothing short of amazing. We are also grateful to Whitney E. Johnson at the University of Washington Press for her assistance with the final manuscript preparations. In addition, we thank our anonymous reviewers, whose detailed feedback and probing questions brought much to the volume as a whole.

We owe special thanks to our teachers and mentors in the Department of Women's, Gender, and Sexuality Studies and throughout Emory University, particularly Lynne Huffer, Pamela Scully, Kimberly Wallace Sanders, Sander Gilman, Rosemarie Garland-Thomson, Holloway Sparks, Elizabeth Wilson, Deboleena Roy, Michael Moon, Joy Ann McDougall, Linda Calloway, and the late Berky Abreu.

We remain grateful beyond measure to the pathbreakers of feminist science studies and queer studies, many of whom are named in these pages. Finally, we thank all of the contributors to this volume, whose scholarship has inspired and enabled us to imagine something like a queer feminist science studies.

All editors contributed equally to this volume.

Queer Feminist Science Studies

Queer Feminist Science Studies

An Introduction

WE LIVE IN A MEDIATED WORLD, ONE IN WHICH SCIENCE AND TECHnology and scientific ways of understanding are deeply imbricated in the production, contestation, and transformation of gender and sexual norms. Concurrently, gender and sexual norms shape scientific methods and truth claims in myriad ways. As many scholars have argued, science is not a "mirror of nature," but rather an interpretive grid through which people narrativize phenomena using the epistemological resources available in particular times and places (Keller 2010). Science is situated knowledge, and as such always partial, dynamic, open to revision, and contested (Haraway 1991; Harding 2008; Latour 1993). As transdisciplinary objects of study, gender and sexuality exemplify the contextual and situated nature of all forms of knowledge production.

At the same time, science and technology have significant effects—at once productive and constraining—on the organization of gender and sexuality at the societal level and on our individual and relational experiences of ourselves as gendered and sex(ualiz)ed beings. While science and technology can open up new possibilities, their dominant manifestations often reproduce and heighten local and global inequalities (Haraway 2008; Spivak 1999), even those that, on their face, promise social advancement and liberation. New reproductive technologies (NRT) provide a good example. While advances in NRT promise new forms of biological reproduction (for example, scientists may, in the near future, be able to create an embryo from the cells of two "males" or two "females"), these technologies are only available to the economically privileged, overwhelmingly rely on the reproductive labor and/or bodies of low-income women, women of color, and women from the Global South, and may reinforce the fetishization of genetic relatedness (Mamo 2007).

In recent decades, science has become a remarkably productive object for queer feminist critical inquiry. Time and again, queer feminist concerns with materiality, biology, power, and subjectivity have led us straight to science as a site for both undoing and remaking our worlds. Our intervention here is at least dual—we want to further open up what counts as both science and science studies *and* to insist on the importance of (traditional and nontraditional) science and science studies to queer feminisms. In addition to arguing for a "queering" of science, science studies, and feminist science studies, we argue that queer feminist approaches to reading science offer profoundly innovative and different answers to some of academic feminism's most formative, trenchant, and enduring questions, including: What is sex? How are race, gender, sexuality, and other systems of difference coconstituted? And how are ideas about and practices of normalization implicated in the maintenance of the status quo?

Queer Feminist Science Studies: A Reader is an eclectic collection of essays that develop, rethink, and expand queer feminist approaches to reading science. Sixteen are previously published and presented in abridged form, and four appear here for the first time. In this reader, we use "queer feminist science studies" to name, nurture, and transform conversations that are already taking place across the sciences, humanities, and social sciences. We offer it as a figurative space where the meanings of and relations among queer and feminist theories and science are reimagined capaciously to foster new critical and creative knowledge-projects. By bringing together intimate but as-yet-unreconciled critical engagements with science, its effects, and/or the "stuff" of the world, we posit queer feminist science studies as a framework that 1) attends to intersecting operations of power and privilege and the material-semiotic makings of normality and deviance, while 2) highlighting potentialities for uncertainty, subversion, transformation, and play.

For this reader, we selected and solicited essays that offer two different types of resources: ones that help us think about science beyond STEM and science studies beyond traditional STS, and ones that highlight the vital import of theorizing science to queer feminisms. Our goal in constructing this collection of queer feminism *as* science studies is to open up questions that haven't been centered in (feminist) science studies as it has been canonized, and to make strange anew those that have. We hope this volume will trouble and enrich narratives of the field of (feminist) science studies, its projects, and thus its genealogies and futures. It is with these goals that we offer this reader as a generative site for forging new understandings not only of the politics of science but also of the dynamic systems (Fausto-Sterling 2012) that give form to what some feminist science scholars have called naturecultural bodies and worlds (Haraway 2008; Barad 2007).

The beauty of the phrase "queer feminist science studies" is perhaps the simultaneity of its material and discursive specificity with its constitutive capaciousness: It grounds and unsettles, offering no fixed archive nor method, but a qualifier—"queer"—that promises to unsettle some of our cherished givens. This could be the first of many volumes offering different archives and new directions for queer forms of feminist science studies. To those ends, it is our hope that readers come away from this volume with a sense of queer feminist science studies as a generative rubric, a useful set of tools, and an intellectual/political space that enables a proliferation of critical and creative responses to processes of naturalization. In queer feminist science studies as we conceive it, naturalization—which is not science itself, but to which science is of course vital—becomes the object of critical and creative interventions. The discourses, objects, and practices with which scholars of queer feminist science studies might engage are vast. The ubiquity of scientific narratives and their power in shaping the parameters of how we think about the distinctions between nature and artifice, biology and culture, time and space, mind and body, and other vital issues means that science studies must roam far beyond institutions of sciences. A queer feminist science studies must concern itself with the quotidian practices of meaning making that trouble, disrupt, and reconfigure assumptions about nature, difference, species, and worldliness. We hope that the volume supports resolve to recoup and reimagine "nature" with eyes wide open to processes of naturalization, both historical and contemporary, and that its readers find the volume rich with resources for holding creative possibility and critical accountability together.

In the remainder of this introduction, we offer further ruminations on queer feminist science studies by tracing its genealogies and describing its frameworks with a particular focus on its contributions to thinking about foundational issues in queer feminism. We also offer an overview of the structure of the reader and of the individual essays and offer some concluding questions.

Definitions (or, What Is Queer Feminist Science Studies?)

What does it mean to undertake a queer feminist study of science? One of the aims of this volume is to open up the word, concept, and practice of "science" (and therefore "science studies") to new, unexpected, and surprising definitions, inhabitations, and hauntings. Science is not a static or unchanging object of analysis. Nor is it the nom de plume of objectivity. Rather, science is dynamic and changing, and it is always political. In addition, disciplinary scientific practices are inextricably conditioned by particular political economies and ideologies that regulate: a) what counts as legitimate scientific knowledge; and b) which kinds of people, bodies, and institutions can conduct scientific research in the first place. Thus, we refuse to equate the practice and theory of science, or that which constitutes the proper object of science studies, with hegemonic scientific practices and paradigms. While many of the essays in this volume do analyze hegemonic scientific practices and paradigms, other essays challenge what "body-knowledges" and materialities (Roy and Subramaniam 2016) count as science and therefore science studies. For example, Jennifer Nash's piece on black anality and Rachel Lee's piece on pussy ballistics take up the discomforting viscera of human life, the awful and the offal, and use them to not only deconstruct but disembowel conventional scientific and feminist models of embodiment alike.

To formulate a queer feminist approach to science studies, we begin with questions familiar to those versed in feminist science studies: Who gets to produce scientific knowledge? Does it matter who produces knowledge and, if so, how (Keller 1983; Harding 1991; Longino 1990)? How do scientific agendas, questions, epistemologies, methods, results, and interpretations reflect gender-normative, racist, heterosexist, classist, and ableist assumptions? How does science justify, create, and enforce social inequalities (Keller 1992; Fausto-Sterling 2000; Haraway 1992b; Schiebinger 1993)? How might we reformulate epistemologies, methods, and interpretations of science in order to do "queer" and or "feminist" science

(Roy 2008; Subramaniam 2014)? How can we use findings about the material-semiotic world (including contingency, variability, and change within "nature") to reformulate our conceptions of the social world (Wilson 2004; Barad 2007)?

To these questions, we add key foci from queer theory. Queer theory has a history of challenging the essentialisms bequeathed by certain legacies of both feminist and lesbian and gay studies to discussions of gender and sexuality. Early queer theory (Foucault 1978; Butler 1990; Sedgwick 1990; Halberstam 1998; Dollimore 1991) troubled assumptions about the natural unity of the category "women" in feminist scholarship on women's ways of knowing, women's time, and the ethics of care (Scott 1988; Kristeva 1981; Held 1993). Departing from these strains of feminist theory (including the conjunction of feminist theory and science studies/history of science and medicine), much 1990s queer theory focused on rethinking debates over the social construction of sexuality (Stein 1990). In the wake of the millennium, queer research returned to the terrain of science studies and deepened understandings of its queer potentialities in the process. For instance, Siobhan Somerville's work on racializing queerness and queering racial science (2000) thoroughly upended the idea that race and sexuality are separate or separable categories in the history of science and medicine. In a similar way, Myra Hird's work on sex/gender variation and sexual diversity in nonhuman animals (2006) refigured the natural sciences as a crucial object of and resource for strengthening queer and trans critique.

In these and other ways, we argue that queer theory sharpens feminist engagements with science by turning critical attention to the imbrication or dependence of center and periphery, inside and outside, natural and artificial, and normativity and deviance. Rather than being opposed, these terms are coentangled divisions produced through specific histories. Thus, there is now a rich array of queer work that contests the naturalization of the categories of normal and deviant sexuality and binarized notions of sexed anatomy, gender identity, sexual desire, and sexual identity (Halberstam 2005; Patton and Sánchez-Eppler 2000; Fausto-Sterling 2000; Downing, Morland, and Sullivan 2015; Gupta 2015; Willey 2016b). Queer frameworks provide useful tools for analyzing the sexual logics of various systems, discourses, and material practices including medicalization (Butler 2004), disease management (Patton 1990), neoliberalism (Duggan

2003; Eng 2010), US imperialism (Puar 2007), and, indeed, subjectivity and sociality themselves (Berlant and Edelman 2013; Edelman 2004; Wilson 2015). Concomitantly, queer theory, like its critical counterparts in gender and sexuality studies, reveals that heterosexist regulatory regimes intersect with structures of race, class, gender, nation, age, and ability to shape life chances in uneven ways (Ferguson 2004; Holland 2012; Grewal and Kaplan 2001; Boellstorff 2005; Arondekar 2009; Stallings 2015). Throughout these instances, queer theory reveals the centrality of science to cultural politics and upends assumptions about nature, progress, and sovereignty.

As scholars interested in pushing the boundaries of both the queer and feminist elements of a queer feminist science studies, we also want to conceptualize queerness in terms that diverge from mainstream feminist theory with regard to the politics of difference, the politics of power, and the politics of privilege. In queer theory, queerness names a structural position of abjection, not a fixed or essential identity, and is therefore irreducible to any single-issue politics or agenda (Spivak 1988; Butler 1993; Cohen 1997; Edelman 2004). In addition, and most importantly perhaps, to queer is to make strange that which seems normal or natural (Sedgwick and Frank 2003), a formulation that is particularly appropriate for the aims of this volume.

The project of queer feminist science studies, we argue, is queer not only in content, but in an approach that seeks to rethink and open up our definitions of science, science studies, and indeed feminism and queer studies too. Bringing these threads together, "queer feminist science studies" might be understood as a way of seeing and foregrounding a shared aim and theoretical/methodological affinity between projects of queer feminism and science studies: what might be described as an ethic of undoing. By *undoing* we mean an intellectual practice of getting underneath a seemingly self-evident idea, to understand the conditions of possibility for its intelligibility. So rather than asking, "How can we represent queer bodies and/or black and brown bodies in more positive ways?" a queer feminist science studies approach animated by an ethic of undoing would ask questions like: How do we know what we know about bodies? What is science? What is nature? What is race? What is sex? What is sexuality? How do these categories intra-act? How do our knowledges of them (epistemology) shape their existence (ontology)?

Contributions of This Framework (or, What Can Queer Feminist Science Studies Do?)

Taken as a whole, the pieces in this volume demonstrate that queer feminist science studies, as we have simultaneously recognized it and called it into being, is an expansive lens, a big basket. It includes diverse feminist and queer analyses of biomedicine, science, technology, and health, as well as critical destabilizations of sexual, gendered, racialized, anthropocentric, and able-bodied logics and hierarchies, among other things. A queer feminist science studies framework calls us to:

1. rethink knowledge production in important ways and challenge entrenched disciplinary divides between the sciences, social sciences, and humanities, thus foregrounding new languages and methodological resources for critical interdisciplinarity (Subramaniam 2014).
2. challenge what "body-knowledges" and materialities (Roy and Subramaniam 2016; Willey 2016a) count as science and, thus, what counts as "science studies."
3. deepen our understandings of onto-epistemological processes & resist the pull to think of knowing and being as separate. As the pieces in this archive show, changes in the world can lead to new ways of thinking, and new ways of thinking can produce material changes in the world, in an "intra-active" and mutually constitutive process (Barad 2007) where the relationship between the ontological (being) and epistemological (knowing) is never very far apart.
4. affirm critique as a site of world-making potential in its own right in the face of calls by some feminist science studies scholars to move beyond it (Coole and Frost 2010; Hird 2004).

These four intimately interconnected charges—to rethink entrenched disciplinarity, expand our archives of materiality, deepen our understandings of the imbrication of knowing and being, and to affirm critique as vital world-making work—are importantly for us always grounded and enabled by a fifth call: to engage debates about power germane to queer feminism. Queer feminist science studies reveals concerns with science, nature, and bodies as vital to central debates within feminist and queer theory, including debates about the sex/gender distinction, the relationship between hierarchies of gender and other hierarchies (such as those of race and ability), and the construction and biopolitical flexibility of categories of normality and deviance. Around each of these nodes, we find thinkers whose projects focus on knowledge politics and whose epistemological preoccupations lend themselves to the productive unmaking of entrenched analytics. These reconfigurations of the terrain of debate are part of this volume, and we offer a few framing remarks on them here.[1]

Sex/Gender

The sex/gender distinction has been taken for granted both in the sciences and in many feminist approaches. Unlike mainstream scientific and (some) feminist approaches, queer feminist science studies does not assume that the sex/gender distinction is either analytically or materiality stable; nor does it presume that this distinction is adequate to critically mapping complicated dimensions of embodiment (Kessler 1998; Fausto-Sterling 2000; Butler 2004; Salamon 2010). Building on this work, and actively figuring *sexing* and *gendering* as verbs and performative processes of materialization, many of the essays in this reader attend to the ways in which queer feminist engagements with science refigure the intra-activity between biology/culture and sex/gender (Fausto-Sterling 2012; Jordan-Young 2010), and thereby provide exciting, capacious, and nondualistic ways of thinking about embodiment. For example, Sarah Richardson's essay on gender bias in sex chromosome research, David Rubin's essay on the genealogy of intersex and the sex/gender distinction in Western biomedicine, Angela Willey and Sara Giordano's essay on sexual dimorphism in monogamy gene research, and Rachel Lee's essay on the biopolitics and geopolitics of racialized gender suggest that sex and gender are neither opposed nor simply mutually constitutive, but are rather multiply interrogational, overdetermined, and entangled with other categories. Investigating these entanglements and intersections, the contributors to this volume develop new models of sexing/gendering bodies that not only complicate and contest but also enrich and transform contemporary scientific, feminist, and queer paradigms, methodologies, and historiographies.

In terms of pedagogy, these essays also suggest new ways of teaching more inclusively about sex/gender. For example, instructors of both introductory and advanced courses on gender and sexuality may introduce intersex and trans to make arguments about other issues, such as the social construction of gender, or the nature and/or value of human diversity. Although well intentioned, this pedagogical approach often has a tokenizing effect. It implies that intersex and trans people's sexes, genders, bodies, and/or subjectivities are somehow more constructed than those of cisgender and nonintersex folks, and naturalizes cisgender and nonintersex lives as normative. The essays we include here, on the other hand, refuse these gestures. They decline to reify intersex, trans, and gender nonconforming subjects into exceptional objects of study, arguing instead that these subjects reveal issues and questions of major or even universal significance. Or, to put it differently, intersex and trans studies are not mere minoritarian concerns, but are rather central to and also critically reconfigure longstanding debates about what it means to be human and what it means to occupy a world with others.

Antiracism and Intersectionality

In regards to theories of and debates surrounding intersectionality (Hill Collins and Bilge 2016), queer feminist science studies focuses on analyzing the conceptual and historical coconstitution of processes of sexing/gendering bodies with race, class, disability, geopolitics, and other social processes in a way that troubles the presumed isometry of sex, gender, and sex/gender as coherent categories unto themselves. Diverse intersections are explored throughout the volume; for example, between gender, sexuality, and disability in Kafer, between gender, sexuality, and nation in Chen and Lee, and between gender, sexuality, and class in Chen and Sullivan.

In particular, one of the underappreciated and undertheorized contributions of and to queer feminist science studies is the substantial work that has been done to theorize the nature of the relationship of concepts and categories of race to those of sex, gender, and sexuality. Some scholars well known in feminist science studies have long been engaged in this project (Schiebinger 1993; Stepan 1993; Haraway 1992a; Hammonds 1994; Somerville 2000; Subramaniam 2009; Irni 2013). The authors in this volume reveal queer feminist science studies as a vital lens for thinking through the historical and material enmeshment of racial/sexual formation. At the heart of our treatment of racial/sexual formation in the curating of this volume is the question: If, as Baradian materialism suggests, we are responsible not only for how we represent the world, but for what exists, how do we understand the onto-epistemological status of race vis-à-vis our materialist feminist engagements with sex, gender, and sexuality?

Pieces such as those by Markowitz, Storr, McWhorter, Nash, Chen, and Thakor suggest that any focus on disrupting (or refining) binary models of sex/gender/sexuality—at the level of molecules or social scripts—risks recentering whiteness by failing to recognize masculinity and femininity as racial designations that have placed bodies not on a continuum per se, but rather on a multidimensional grid of raced/gendered/sexed types. As Sally Markowitz argues in this volume, and captured beautifully in Moya Bailey's (2016) concept of *misogynoir*, when we invoke an unmarked gender binary as a core problematic of feminist theory, we risk recentering whiteness, because gender was never "just two," but an always racialized and pluralized scientific concept.

We hope that the essays in this reader will provoke questions about how to better (and perhaps more systematically) approach thinking about the historical and material enmeshment of these categories. We also hope it will serve as a resource for folks to consider more concretely the racial implications of the operationalization of sex/gender/sexuality as objects of knowledge across the natural sciences, social sciences, and humanities. And finally, we hope that this reader will recenter the study of racial formations as a radically interdisciplinary project, not one that belongs strictly to "culture," but one at the heart of the naturecultural world we hope to know and make differently (Holland 2012; Wynter 2003; Spillers 2003).

On Normalization

Influenced by the work of Michel Foucault, queer feminist science studies approaches insist that the construction of normality and deviance is a material-semiotic process. At the center of many of the essays in this volume are the following questions: How are categories of normality/health and deviance/pathology in relation to sex, gender, sexuality, race, ability, class, and nation created and reproduced across populations? How do these norms become embodied at an individual level? How are they transformed at a societal level? How are they resisted at an individual level?

There is a rich tradition of asking these questions in feminist scholarship, queer and sexuality studies, and disability studies. A classic example of this scholarship is Anne Koedt's "Myth of the Vaginal Orgasm" (1970), in which she argues that "scientific authorities" have propagated and enforced categories of normal and deviant female sexuality (orgasmic vaginality vs. anorgasmia or orgasmic clitorality) through the diagnosis and treatment of "frigidity." A number of scholars, including Emily Martin (2001), Dorothy Roberts (1997), Adele Clarke (1998), Laura Briggs (2002), Susan Bordo (1993), and many more, have built on the work of Koedt. Similarly, there is a rich tradition within queer studies, trans studies, and intersex studies of exploring the construction of normality and deviance in the response to the HIV/AIDS epidemic (Epstein 1996; Treichler 1999; Crimp 2002; Patton 1990), in the pathologization of homosexuality and gender nonconformity (Bayer 1981; Stein 1999; Terry 1999; Rosario 2002; Carter 2007; Meyerowitz 2004; Najmabadi 2013; Spade 2003), and in the medical response to intersex infants (Kessler 1998). More recently, disability studies scholars have offered further insight into the construction of normality and deviance through the medicalization of nonnormative bodies.

Building on this work, essays in this volume such as those by McWhorter, Kafer, Race, Sullivan, and Chen suggest that queer feminist science studies approaches understand the construction of normality and deviance as a material-semiotic, naturecultural process. Nikki Sullivan argues, in regard to the case she analyzes, that "two-handedness" was constructed as normal and "one-handedness" as deviant through media labeling of the desire for hand-amputation as mentally disordered and through the refusal of doctors to remove a transplanted hand that was no longer wanted (and that both of these were material-semiotic processes). Queer feminist science studies also assumes that resistance to regimes of normality and deviance occurs through material-semiotic processes, such as when Michelle Murphy argues that women's health activists challenged ideas about normal and deviant cervixes by meticulously recording data about their own cervixes every day for months and then using this data to argue that there is not any one model of cervical health. Queer feminist sciences studies scholarship offers powerful examples of how to critique the normalizing and pathologizing tendencies of Western biomedicine and science, while also providing insights into how science and biomedicine might be reformed and/or (re)appropriated in order to better serve the interests of women, sexual minorities, gender-nonconforming people, and other historically marginalized groups. In general, queer feminist science studies of normality and deviance offer important resources for thinking about questions of normalization, resistance and transformation, structure and agency.

Partial Genealogies (or, Where Did Queer Feminist Science Studies Come From?)

We became convinced of the need for a volume on queer feminist science studies when, individually and collectively, we were asked questions about whether and how science studies might be "queered" and about whether and how queer feminism might be transformed by new understandings of biology, species, and matter. These questions were exciting, but also somewhat confounding, because there is already such a rich archive of, well, queer feminist science studies. But this work is not necessarily read as such, and queerness has not always been centered within articulations of the field of feminist science studies. Thus, we argue that, somewhat paradoxically, queer feminist science studies is both new and old. It is new because it is a new lens, a new basket for collecting together scholarship that has not necessarily been grouped together before. And it is old in that the basket is not only shaped by long histories but also capable of holding scholarship from throughout the history of queer feminist thought.

Our shared training and diverse interests shape the properties of this metaphorical basket. We come to this project as colleagues who have been in dialogue about sometimes disparate and sometimes synergetic interests in the biosciences for about a decade. The four editors of this volume were trained in the Department of Women's, Gender, and Sexuality Studies at Emory University in the first decade of the twenty-first century. Unlike feminist science studies scholars of previous generations, most of whom were trained in traditional disciplines (in STEM, the social sciences, and humanities) and then migrated to women's studies, each of us came to conceptualize the critical study of science as central to our work through our training in the interdisciplinary study of women, gender, and sexuality and, in particular, the transdisciplinary tools of queer and feminist theory. Thus, although our understanding of queer feminist science studies comes to us from feminist science studies and queer studies primarily, it is undergirded by theory from literary studies, intersex and transgender studies, cultural studies, disability studies, feminist studies of the body/embodiment, ecofeminism, postcolonial (science) studies, reproductive justice movements, queer of color critique, feminist ontology, feminist theology, animal studies, feminist speculative fiction, science fiction (sci-fi) and sci-art of all forms, technology and digital media studies, performance studies, and studies of science and medicine in other disciplines. Many of the most influential and radical works from these areas are unnamed in this volume, although its pages bear their legacies. We owe much, for example, to Hélène Cixous's and Mary Daly's new languages of natural order, Hortense Spillers's semiotics of race and sex, Orlan's unflinching articulations of the hybrid body, and the viscerality of Audre Lorde's auto-bio-mythographic refrains. And we cannot forget the debts we owe to those who cannot be named: those who found themselves subject to brutal and unforgiving scientific scrutiny, those forced to sit by while their work was discredited or misattributed, and those who marched, collectively and individually, for the recognition of their own (scientific, medical, cultural) integrity.

We want the basket of queer feminist science studies to hold together scholarship borne across intellectual traditions and disciplines. The particular collection presented here comes from diverse places, including feminist science studies "proper" (e.g., Sarah Richardson, "Sexing the X: How the X Became the 'Female Chromosome'") feminist disability studies (e.g., Alison Kafer, "At the Same Time, Out of Time: Ashley X") somatechnics (e.g., Nikki Sullivan, "'BIID'? Queer [Dis]Orientations and the Phenomenology of 'Home'"), and queer/critical public health (e.g., Kane Race, "Embodiments of Safety"). Few of the previously published pieces in this volume were written using the framework of "queer feminist science studies" explicitly, and resituating these works is a way of recruiting fellow travelers into our current intellectual and political project(s). By contrast, the four essays written especially for this volume use the framework of "queer feminist science studies" more intentionally, giving examples of what new work is made possible within this new lens.

As editors, we prioritized recent scholarship, essays from emerging scholars, and essays that we saw as underrecognized and ripe to be reckoned with. Thus, a cadre of scholars whose work has been vital to feminist science studies field formation and who have been doing "queer feminist science studies" for decades don't appear here. Evelyn Fox Keller's groundbreaking work, including early work on slime molds; Sandra Harding's work on starting from lesbian lives; Anne Fausto-Sterling's work on the sexing of genitals, brains, hormones, and skeletons; Evelynn Hammonds's work on racialization and sexual politics; Donna Haraway's insistence, for decades, on making strange what we think we know; Dorothy Roberts's work on reproductive justice; Jennifer Terry's work on the biologization of sexuality and the study of queerness in human and nonhuman animals; and Cathy Cohen's work on HIV and AIDS are all major works of scholarship to which we, and the field at large, owe an enormous debt of gratitude—both for the questions they centered and the conversations they started.

In addition, because one of our aims in this volume is to show that queer feminist readings of science have been at the center of certain academic feminist debates, there are notable absences in this volume. This volume consists primarily of scholarship by academics who write and teach in (but not exclusively about) the Global North and who were trained in humanistic and social scientific disciplines. We foresee collections of queer feminist science studies that highlight queer feminist interdisciplinarity and decenter Western thought. There is much material for a volume that highlights groundbreaking work in queer feminist speculative fictions, both as Aimee Bahng has argued (in this volume) and in the legacy of feminist science fiction and cyberpunk readers (Little 2007; Flanagan and Booth 2002; Imarisha and brown 2015; Larbalestier 2006), or a volume showcasing sites of scientific research through queer feminist lenses or queer and feminist materialist classrooms (Subramaniam 2014; Smith, Meyers, and Cook 2014; Giordano 2016). Finally, there is a curious bias in feminist science studies toward US-centered topics and debates, despite important exceptions and despite the internationalism of science studies more broadly (Subramaniam 2009). As we constructed this reader, we felt it was important to include both types of essays: ones that rethink conventional US histories of science in queer feminist ways; and also ones that critique and challenge the unmarked US-centrism of feminist and queer approaches to science. Essays by Thakor, Lee, Sullivan, and Chen variously theorize and contest the national and transnational relations of power and knowledge that overdetermine normative versus deviant bodies and practices in and across different cultural and scientific contexts. Ultimately, we hope that readers will extract theoretical gains from these essays that enable them to address this and other critical lacunae in (queer) feminist science studies in future work.

This volume suggests that as we aspire to "queer" (feminist) science studies or to "scientize" queer feminism, we have rich traditions and scholarship on which to draw and to which we might productively see ourselves as accountable. It is our hope that in naming queer feminist science studies as a particular field, subfield, framework, and/or project, potentialities for undisciplining and repoliticizing these questions will proliferate.

Organization and Contents (or, What Will You Find in This Reader, Anyway?)

This volume is organized topically into four parts. Each part is preceded by an introduction highlighting connections and points for discussion, so here we only introduce the parts briefly and conclude with some questions raised by the connections between them. The first part, "Histories of Difference," investigates the processes through which different differences have come to matter within the history of science and thus historicizes the naturalized status quo. Part One includes a piece by Sarah S. Richardson on the feminization of the X chromosome in scientific and popular discourse, a piece by Sally Markowitz on the racialization of gender categories in the history of science, a piece by Merl Storr on the racialized genealogy of the term *bisexuality* in sexology and psychoanalysis, a piece by Ladelle McWhorter on the role of scientific racism and notions of sexual predation in shaping dominant twentieth-century figurations of homosexuality, and a piece by David Rubin on the role of sexological approaches to intersex people in producing the category of gender in the later half of the twentieth century.

The second part, "Contemporary Archives and Case Studies," comprises a series of readings of specific archives and medical cases in light of the histories examined in Part One, offering a set of critical science literacy skills that can be applied to contemporary objects of analysis. Part Two includes a piece by Jennifer Nash on representations of black female sexuality in a "nonscientific" discourse about human anatomy and physiology—namely pornography; a piece by Alison Kafer on the ableism and the erasure of crip desire involved in the treatment of "Ashley X," a girl with significant cognitive disabilities; a piece by Nikki Sullivan on "Body Integrity Identity Disorder" as a "somatechnology" that serves to shore up the boundaries of the normative subject; a piece by Mitali Thakor on the racial/sexual assumptions undergirding "Project Sweetie," an anti-trafficking digital community policing project; and a piece by Hilary Malatino on the raced and classed absences of intersex and trans medical/scientific archives.

The third part, "Disruptive Practices," comprises a series of pieces that analyze and/or imagine practices aimed at disrupting the status quo. Part Three includes a piece by Michelle Murphy on the efforts of feminist women's health activists to disrupt normative understandings of cervical health; a piece by Rachel Lee on comedian Margaret Cho's use of peristaltic feminism to disrupt Western imperialism; a piece by Kane Race imagining a disruptive HIV/AIDS harm reduction strategy focused on pleasure and embodiment; a piece by Amber Musser imagining ways of thinking about sexual consent that disrupt legal demands for "able-mindedness"; and a piece by Isabelle Dussauge imagining a utopic "queer neuroscience."

The final part, "Beyond the Human," explores how queer feminist science studies scholars interrogate the boundaries of humanity in ways that open narrative possibilities for knowing and becoming otherwise. Part Four includes a piece by Angela Willey and Sara Giordano on intertwining of sex/gender and sexuality in monogamy research on voles, a piece by Vicki Kirby proposing a rearticulation of human/nonhuman via nature/culture, selections from Luciana Parisi's *Abstract Sex* in which she considers the queer potentials of sex without bodies, a piece by Mel Chen examining toxicity and the inanimate, and a piece by Aimee Bahng considering Octavia Butler's speculative fabulation as a practice of queer feminist science studies.

Taken together, these essays encourage us to confront questions vital to the articulation of queer feminist science studies: Which differences matter, when, and where? What stands in for "the body," what are bodies, what matters about bodies, to whom, and why? How are processes of naturalization and normalization intertwined, and what does it mean to engage processes of medicalization ethically in light of histories of eugenics and ongoing struggles for access to the fruits of science and technology? What does it look like to move beyond anthropocentric thinking in worlds where "humanity" is distributed unevenly among humans?

Looking Forward (Assuming Time Is Linear)

Where might queer feminist science studies go from here? For the editors of and contributors to this volume, this is necessarily an open question. This openness to the unexpected, the uncertain, and the unknowable is perhaps a defining feature—and we believe one of the most generative contributions—of queer feminist science studies to reimagining science, science studies, women's, gender, and sexuality studies, and interdisciplinary scholarship more broadly. "It is the task of the imagination," Gayatri Chakravorty Spivak writes in her most recent book *Readings*, "to place a question mark upon the declarative" (Spivak and Choksey 2014, 4). For Spivak, a literary critic, literature is the privileged location of the imaginary and imaginative thinking. But we contend that science and queer feminism too—in their many instantiations—manifest underanalyzed yet vital question marks upon the ineffably interconnected and contested declaratives of the present, past, and future. Closely related to what Spivak earlier called "persistent critique" (1993), queer feminist science studies in particular is reimaginative precisely in its commitment to thinking difference differently, its healthy appreciation for onto-epistemological disobedience to normative regimes, and its emphasis on the ethico-political stakes of our relationships to innumerable others, near and afar, who make our own lives, communities, and works possible.

For our part, we hope that this volume demonstrates the diversity and richness of queer feminist critical inquiries into science and presents a range of possible directions for future study. We hope this volume offers stronger and queerer feminist genealogies for feminist new materialism, feminist science and science studies, critical disability studies, and critical studies of gender, sexuality, and embodiment in their myriad manifestations. We hope those genealogical resources will ground us as we explore evolving sciences, and the stuff of our worlds, in the knowledge that contexts of intelligibility can never be relegated to the footnotes. The world itself is material-discursive; there is no biology without history, no materiality that precedes power. And we hope that queer feminisms broadly conceived will find in the volume science—as a site of critique, reclamation, and imagination—reflected back to itself as vital to its world-making work.

Notes

1 In fact, in our initial conceptualization of the reader, we had envisioned a section on each of these themes. After consultation with our reviewers and editors, we realized that these themes run through all of the pieces in the reader and reorganized accordingly.

Works Cited

Arondekar, Anjali R. 2009. *For the Record: On Sexuality and the Colonial Archive in India*. Durham: Duke University Press.

Bailey, Moya. 2016. "Misogynoir in Medical Media: On Caster Semenya and R. Kelly." *Catalyst: Feminism, Theory, Technoscience* 2 (2).

Barad, Karen Michelle. 2007. *Meeting the Universe Halfway: Quantum Physics and the Entanglement of Matter and Meaning*. Durham: Duke University Press.

Bayer, Ronald. 1981. *Homosexuality and American Psychiatry: The Politics of Diagnosis*. New York: Basic Books.

Berlant, Lauren Gail, and Lee Edelman. 2013. *Sex, or the Unbearable*. Durham: Duke University Press.

Boellstorff, Tom. 2005. *The Gay Archipelago: Sexuality and Nation in Indonesia*. Princeton: Princeton University Press.

Bordo, Susan. 1993. *Unbearable Weight: Feminism, Western Culture, and the Body*. Berkeley: University of California Press.

Briggs, Laura. 2002. *Reproducing Empire: Race, Sex, Science, and U.S. Imperialism in Puerto Rico*. Berkeley: University of California Press.

Butler, Judith. 1990. *Gender Trouble: Feminism and the Subversion of Identity*. East Sussex: Psychology Press.

———.1993. *Bodies That Matter: On the Discursive Limits of "Sex."* New York: Routledge.

———. 2004. *Precarious Life: The Powers of Mourning and Violence*. New York: Verso.

Carter, Julian B. 2007. *The Heart of Whiteness: Normal Sexuality and Race in America, 1880–1940*. Durham: Duke University Press.

Clarke, Adele. 1998. *Disciplining Reproduction: Modernity, American Life Sciences, and "The Problems of Sex."* Berkeley: University of California Press.

Cohen, Cathy J. 1997. "Punks, Bulldaggers, and Welfare Queens: The Radical Potential of Queer Politics?" *GLQ* 3: 437–465.

Coole, Diana H., and Samantha Frost. 2010. *New Materialisms: Ontology, Agency, and Politics*. Durham: Duke University Press.
Crimp, Douglas. 2002. *Melancholia and Moralism: Essays on AIDS and Queer Politics*. Cambridge, MA: MIT Press.
Dollimore, Jonathan. 1991. *Sexual Dissidence: Augustine to Wilde, Freud to Foucault*. Oxford: Oxford University Press.
Downing, Lisa, Iain Morland, and Nikki Sullivan. 2015. *Fuckology: Critical Essays on John Money's Diagnostic Concepts*. Chicago: University of Chicago Press.
Duggan, Lisa. 2003. *The Twilight of Equality?: Neoliberalism, Cultural Politics, and the Attack on Democracy*. Boston: Beacon Press.
Edelman, Lee. 2004. *No Future: Queer Theory and the Death Drive*. Durham: Duke University Press.
Eng, David L. 2010. *The Feeling of Kinship: Queer Liberalism and the Racialization of Intimacy*. Durham: Duke University Press.
Epstein, Steven. 1996. *Impure Science: AIDS, Activism, and the Politics of Knowledge*. Berkeley: University of California Press.
Fausto-Sterling, Anne. 2000. *Sexing the Body: Gender Politics and the Construction of Sexuality*. 1st ed. New York: Basic Books.
———. 2012. *Sex/Gender: Biology in a Social World*. New York: Routledge.
Ferguson, Roderick A. 2004. *Aberrations in Black: Toward a Queer of Color Critique*. Minneapolis: University of Minnesota Press.
Flanagan, Mary, and Austin Booth. 2002. *Reload: Rethinking Women + Cyberculture*. Cambridge, MA: MIT Press.
Foucault, Michel. 1978. *The History of Sexuality, Vol 1: An Introduction*. 1st American ed. New York: Pantheon Books.
Giordano, Sara. 2016. "Building New Bioethical Practices through Feminist Pedagogies." *International Journal of Feminist Approaches to Bioethics* 9 (1): 81–103.
Grewal, Inderpal, and Caren Kaplan. 2001. "Global Identities: Theorizing Transnational Studies of Sexuality." *GLQ: A Journal of Lesbian and Gay Studies* 7 (4): 663–679.
Gupta, Kristina. 2015. "Compulsory Sexuality: Evaluating an Emerging Concept." *Signs: Journal of Women in Culture and Society* 41 (1): 131–154.
Halberstam, Judith. 1998. *Female Masculinity*. Durham: Duke University Press.
———. 2005. *In a Queer Time and Place: Transgender Bodies, Subcultural Lives*. New York: NYU Press.
Hammonds, Evelynn. 1994. "Black (W)holes and the Geometry of Black Female Sexuality." *differences: A Journal of Feminist Cultural Studies* 6 (2–3): 126–146.
Haraway, Donna. 1991. *Simians, Cyborgs, and Women: The Reinvention of Nature*. New York: Routledge.
———. 1992a. *Primate Visions: Gender, Race, and Nature in the World of Modern Science*. London: Verso.
———. 1992b. "The Promise of Monsters: A Regenerative Politics for Inappropriate/d Others." In *Cultural Studies*, edited by Lawrence Grosberg, Cary Nelson, and Nelson Treichler, 295–337. New York: Routledge.

———. 2008. *When Species Meet*. Minneapolis: University of Minnesota Press.
Harding, Sandra G. 1991. *Whose Science? Whose Knowledge?: Thinking from Women's Lives*. Ithaca: Cornell University Press.
———. 2008. *Sciences from Below: Feminisms, Postcolonialities, and Modernities*. Durham: Duke University Press.
Held, Virginia. 1993. *Feminist Morality: Transforming Culture, Society, and Politics*. Chicago: University of Chicago Press.
Hill Collins, Patricia, and Sirma Bilge. 2016. *Intersectionality*. Cambridge: Polity.
Hird, Myra J. 2004. *Sex, Gender, and Science*. New York: Palgrave Macmillan.
———. 2006. "Animal Transex." *Australian Feminist Studies* 21 (49): 35–50.
Holland, Sharon Patricia. 2012. *The Erotic Life of Racism*. Durham: Duke University Press.
Imarisha, Walidah, and adrienne maree brown, eds. 2015. *Octavia's Brood: Science Fiction Stories from Social Justice Movements*. Chico: AK Press.
Irni, Sari. 2013. "The Politics of Materiality: Affective Encounters in a Transdisciplinary Debate." *European Journal of Women's Studies* 20 (4): 347–360.
Jordan-Young, Rebecca. 2010. *Brain Storm: The Flaws in the Science of Sex Differences*. Cambridge, MA: Harvard University Press.
Keller, Evelyn Fox. 1983. *A Feeling for the Organism: The Life and Work of Barbara McClintock*. San Francisco: W.H. Freeman.
———. 1992. *Secrets of Life, Secrets of Death: Essays on Language, Gender, and Science*. New York: Routledge.
———. 2010. *The Mirage of a Space between Nature and Nurture*. Durham: Duke University Press.
Kessler, Suzanne J. 1998. *Lessons from the Intersexed*. New Brunswick: Rutgers University Press.
Koedt, Anne. 1970. "The Myth of the Vaginal Orgasm." *Radical Feminism: A Documentary Reader*, edited by Barbara A. Crow, 371–377. New York: NYU Press.
Kristeva, Julia. 1981. "Women's Time." *Signs: Journal of Women in Culture and Society* 7 (1): 13–35.
Larbalestier, Justine. 2006. *Daughters of Earth: Feminist Science Fiction in the Twentieth Century*. Middletown: Wesleyan University Press.
Latour, Bruno. 1993. *We Have Never Been Modern*. Cambridge, MA: Harvard University Press.
Little, Judith A. 2007. *Feminist Philosophy and Science Fiction: Utopias and Dystopias*. Amherst: Prometheus Books.
Longino, Helen E. 1990. *Science as Social Knowledge: Values and Objectivity in Scientific Inquiry*. Princeton: Princeton University Press.
Mamo, Laura. 2007. *Queering Reproduction: Achieving Pregnancy in the Age of Technoscience*. Durham: Duke University Press.
Martin, Emily. 2001. *The Woman in the Body: A Cultural Analysis of Reproduction*. Boston: Beacon Press.
Meyerowitz, Joanne J. 2004. *How Sex Changed: A History of Transsexuality in the United States*. 1st paperback ed. Cambridge, MA: Harvard University Press.

Najmabadi, Afsaneh. 2013. *Professing Selves: Transsexuality and Same-Sex Desire in Contemporary Iran*. Durham: Duke University Press.

Patton, Cindy. 1990. *Inventing AIDS*. New York: Routledge.

Patton, Cindy, and Benigno Sánchez-Eppler. 2000. *Queer Diasporas*. Durham: Duke University Press.

Puar, Jasbir K. 2007. *Terrorist Assemblages: Homonationalism in Queer Times*. Durham: Duke University Press.

Roberts, Dorothy E. 1997. *Killing the Black Body: Race, Reproduction, and the Meaning of Liberty*. 1st ed. New York: Pantheon Books.

Rosario, Vernon A. 2002. *Homosexuality and Science: A Guide to the Debates*. Santa Barbara: ABC-CLIO.

Roy, Deboleena. 2008. "Asking Different Questions: Feminist Practices for the Natural Sciences." *Hypatia* 23 (4): 134–156.

Roy, Deboleena, and Banu Subramaniam. 2016. "Matter in the Shadows: Feminist New Materialism and the Practices of Colonialism." In *Mattering: Feminism, Science, and Materialism*, edited by Victoria Pitts-Taylor, 23–42. New York: NYU Press.

Salamon, Gayle. 2010. *Assuming a Body: Transgender and Rhetorics of Materiality*. New York: Columbia University Press.

Schiebinger, Londa L. 1993. *Nature's Body: Gender in the Making of Modern Science*. Boston: Beacon Press.

Scott, Joan Wallach. 1988. *Gender and the Politics of History*. New York: Columbia University Press.

Sedgwick, Eve Kosofsky. 1990. *Epistemology of the Closet*. Berkeley: University of California Press.

Sedgwick, Eve Kosofsky, and Adam Frank. 2003. *Touching Feeling: Affect, Pedagogy, Performativity*. Durham: Duke University Press.

Smith, Pamela H., Amy R. W. Meyers, and Harold J. Cook. 2014. *Ways of Making and Knowing: The Material Culture of Empirical Knowledge*. Ann Arbor: University of Michigan Press.

Somerville, Siobhan B. 2000. *Queering the Color Line: Race and the Invention of Homosexuality in American Culture*. Durham: Duke University Press.

Spade, Dean. 2003. "Resisting Medicine, Re/modeling Gender." *Berkeley Women's Law Journal* 18 (1): 15–39.

Spillers, Hortense J. 2003. *Black, White, and in Color: Essays on American Literature and Culture*. Chicago: University of Chicago Press.

Spivak, Gayatri Chakravorty. 1988. "Can the Subaltern Speak?" In *Marxism and the Interpretation of Culture*, edited by Cary Nelson and Lawrence Grossberg, 271–313. Urbana: University of Illinois Press.

———. 1993. *Outside in the Teaching Machine*. New York: Routledge.

———. 1999. *A Critique of Postcolonial Reason: Toward a History of the Vanishing Present*. Cambridge, MA: Harvard University Press.

Spivak, Gayatri Chakravorty, and Lara Choksey. 2014. *Readings*. London: Seagull.

Stallings, LaMonda Horton. 2015. *Funk the Erotic: Transaesthetics and Black Sexual Cultures*. Chicago: University of Illinois Press.
Stein, Edward. 1990. *Forms of Desire: Sexual Orientation and the Social Constructionist Controversy*. New York: Garland.
———. 1999. *The Mismeasure of Desire: The Science, Theory and Ethics of Sexual Orientation*. Oxford: Oxford University Press.
Stepan, Nancy. 1993. "Race and Gender: The Role of Analogy in Science." In *The "Racial" Economy of Science: Toward a Democratic Future*, edited by Sandra G. Harding, 359–76. Bloomington: Indiana University Press.
Subramaniam, Banu. 2009. "Moored Metamorphoses: A Retrospective Essay on Feminist Science Studies." *Signs: Journal of Women in Culture and Society* 34 (4): 951–980.
———. 2014. *Ghost Stories for Darwin: The Science of Variation and the Politics of Diversity*. Urbana-Champaign: University of Illinois Press.
Terry, Jennifer. 1999. *An American Obsession: Science, Medicine, and Homosexuality in Modern Society*. Chicago: University of Chicago Press.
Treichler, Paula A. 1999. *How to Have Theory in an Epidemic: Cultural Chronicles of AIDS*. Durham: Duke University Press.
Willey, Angela. 2016a. "Biopossibility: A Queer Feminist Materialist Science Studies Manifesto, with Special Reference to the Question of Monogamous Behavior." *Signs: Journal of Women in Culture and Society* 41 (3): 553–577.
———. 2016b. *Undoing Monogamy: The Politics of Science and the Possibilities of Biology*. Durham: Duke University Press.
Wilson, Elizabeth A. 2004. *Psychosomatic: Feminism and the Neurological Body*. Durham: Duke University Press.
———. 2015. *Gut Feminism*. Durham: Duke University Press.
Wynter, Sylvia. 2003. "Unsettling the Coloniality of Being/Power/Truth/Freedom: Towards the Human, after Man, Its Overrepresentation—An Argument." *CR: The New Centennial Review* 3 (3): 257–337.

PART ONE

Histories of Difference

THIS ANTHOLOGY OPENS WITH A SET OF ESSAYS THAT ANALYZE AND contextualize what we call, drawing on Jill A. Fisher (2011), histories of difference in the sciences. Scientific studies of human difference—in particular, differences of sex, gender, race, sexuality, age, and ability—are historically contingent and contested, as are the differences being studied and codified under these labels themselves. The genealogies of specific disciplines and objects included in this section highlight this contingency and reveal key insights into the systematic and white supremacist power relations that normalize certain conceptions and configurations of difference. History, ideology, and culture shape scientific research on differences among human populations and bodies in ways large and small. Because sciences are dynamic and change over time, the consequences—at once material and semiotic—of the enmeshment of science in history and culture cannot be predicted in advance or once and for all. We need, in other words, to constantly forge new analytic lenses and new forms of critical science literacy (Giordano, forthcoming).

This section suggests that contemporary scientific approaches to difference must be historicized in relation to earlier scientific efforts to classify human types—namely, eugenics, sexology, and related bioscientific fields. Classically defined, eugenics promotes higher rates of reproduction for populations with genetic traits considered desirable and reduced rates of reproduction for populations marked as biologically and/or culturally inferior. Sexology, the scientific study of human sex development, developed contemporaneously and in conversation with eugenics, and played a formative role in the production and naturalization of the normal/pathological distinction (Downing, Morland, and Sullivan 2015). The origins of eugenic and sexological thinking are of heterogeneous provenance, dating as far back as ancient Greece, but their modern forms can be traced to European imperial and American settler colonial policies and their adjacent scientific elaborations and justifications in the eighteenth, nineteenth, and twentieth centuries. Feminist and queer scholars have critiqued eugenics and sexology for upholding Eurocentric, white supremacist, heteropatriarchal, compulsory able-bodied, and biological determinist presuppositions. They have also analyzed the rise and circulations of neo-eugenic and neosexological ideas in scientific research and policymaking related to population control, reproductive rights, welfare, the AIDS crisis, the obesity epidemic, sexual health and education, disability rights, environmental degradation, indigenous sovereignty, assisted reproductive technologies, and numerous other areas of contemporary concern.

As these examples suggest, sciences of human differences constitute a useful site for interrogating the intersectionality of race, class, sex, gender, sexuality, age, and ability. As Banu Subramaniam astutely surmised in "Moored Metamorphoses: A Retrospective Essay on Feminist Science Studies" (2009), feminist scholarship needs science studies to help us understand what "intersectionality" is. Conversely, queer and feminist engagements toward this end should be recognized as an important resource for addressing some of feminist science studies' most pressing (and queerest) questions. What counts as difference in evolutionary theory, physiology, biochemistry, endocrinology, genetics, and neuroscience? What is being measured and how? How do sciences of difference in turn shape biomedical, biopolitical, and geopolitical practices? Taken as a whole, the essays in Part One suggest that histories of difference deserve sustained femi-queer genealogical excavation and critical analysis precisely because of their lasting impact on how we come to value, study, treat, manage, and respond to variation (Subramaniam 2014).

In the opening excerpted piece in this section, "Sexing the X: How the X Became the 'Female Chromosome,'" originally published in 2012 in *Signs*, Sarah S. Richardson examines how the X became the "female chromosome" and how the X is gendered female in scientific and popular discourse. The sexing of the X, Richardson argues, represents a case of gender-ideological bias in scientific research, both historically and in the present day. Following in the tradition of Emily Martin (1991), Nelly Oudshoorn (1994), Anne Fausto-Sterling (2000), and Donna Haraway (1997), Richardson shows that feminist science studies disarms and displaces—or queers—accepted understandings of the biochemistry and genetics of sex.

In "Pelvic Politics: Sexual Dimorphism and Racial Difference," first published in *Signs* in 2001, Sally Markowitz makes an intervention into feminist theories of sex and gender as binary systems, arguing that their central conceits have by and large been institutionalized while the implications of the histories out of which they emerge have yet to be carefully reckoned with. She traces the concept of sexual dimorphism—the difference/distance between male and female of a given racial or species type—as a marker of evolutionary development to show that sex "itself" was never actually imagined as binary, but rather always racialized. Racial groups were coded more feminine or masculine, and those with the greatest imagined disparity between sexed bodies, superior. The whiteness of binary sex, a central preoccupation of bioscientific and feminist research, raises important questions for how we understand the meaning of race for femiqueer scholarship and how we figure what we know as "sex" in new biocultural treatments of the materiality of bodies (Richardson 2012; Fausto-Sterling 2000; Grosz 1994).

Merl Storr's "Sexual Reproduction of 'Race': Bisexuality, History and Racialization" is from a 1997 bisexuality studies anthology, *The Bisexual Imaginary: Representation, Identity and Desire*. Like Richardson and Markowitz, Storr highlights the trouble with our (queer) preoccupation with binaries. She reads discourses around the universality of "bisexual" desire through a genealogy of the term. In so doing, the piece shows how the intelligibility of romantic gestures to a sexually free past depends upon the historic racialization of concepts of development in sexology and psychoanalysis. Storr's genealogy suggests that questions of sexuality's materiality must answer to the racial history of its intelligibility as such and prefigures some of the most prescient recent critiques of new materialism, posthumanism, and queer theory. She points to slippages between uses of "race" to mean human, white, European, and British, and thus to the importance of careful attendance to the racial resonances of our claims about categories of human and nonhuman (Weheliye 2014; McKittrick 2014). Storr insists that strategies for resistance to sexual hegemonies should look to history and be accountable to their own conditions of intelligibility (Huffer 2013).

Ladelle McWhorter picks up these threads of connection in a thoroughgoing rethinking of the history of sexuality. Whereas Storr traces "bisexuality," Somerville (2000) "homosexuality," and Willey (2016) "monogamy," through histories of scientific racism, McWhorter, like Ann Stoler (1995), turns to the historiography of sexuality to which these figurations belong. Within the larger book project from which this excerpt, "From Masturbator to Homosexual: The Construction of the Sex Pervert," is drawn, McWhorter offers a genealogical account of "sex" and "race." In the larger book project, she shows how white supremacy shaped conceptions of sexual threat that informed dominant twentieth-century figurations of homosexuality. McWhorter's careful, systematic elaboration of Foucault's history of sexuality through histories of scientific racism is a tremendous resource for queer feminist science studies. By tracking the coformation of various figures of sexual predation, she illustrates the importance of white supremacy to the production of contemporary queer and trans subjectivities with clarity.

David Rubin's "'An Unnamed Blank That Craved a Name': A Genealogy of Intersex as Gender," first published in *Signs* in 2012, addresses the place of intersex bodies in genealogies of the sex/gender distinction. Closely reading the contradictions and legacies of psychoendocrinologist John Money's founding paradigm of intersex treatment, Rubin argues that intersexuality played a crucial role in the invention of gender as a category in mid-twentieth-century biomedical and, subsequently, feminist discourses; that Money used the concept of gender to cover over and displace the biological instability of the body he discovered through his research on intersex; and finally, that Money's conception of gender produced new technologies of psychosomatic normalization. In staging this argument, Rubin reorients genealogies of gender in science by demonstrating the centrality of the medical normalization of intersex subjects to modern understandings of sex/gender and embodiment more generally.

Discussion Questions

1. What are some common threads between the different histories of difference analyzed in Part One? Where do their archives (which sciences and scientists and geographical contexts they study) overlap and diverge?
2. Collectively, the essays in Part One make a strong case for the import of intersectionality to queer feminist genealogies of science. How do these genealogies help us to understand intersectionality? What are the most salient contributions of this approach, and what are its limitations?

Works Cited

Davidson, Phoebe, ed. 1997. *The Bisexual Imaginary: Representation, Identity and Desire*. London: Cassell.

Downing, Lisa, Iain Morland, and Nikki Sullivan. 2015. *Fuckology: Critical Essays on John Money's Diagnostic Concepts*. Chicago: University of Chicago Press.

Fausto-Sterling, Anne. 2000. *Sexing the Body: Gender Politics and the Construction of Sexuality*. 1st ed. New York: Basic Books.

Fisher, Jill A. 2011. *Gender and the Science of Difference: Cultural Politics of Contemporary Science and Medicine*. New Brunswick: Rutgers University Press.

Giordano, Sara. Forthcoming. *The Politics and Ethics of "Labs of Our Own": Post/feminist Tinkerings with Science*.

Grosz, E. A. 1994. *Volatile Bodies: Toward a Corporeal Feminism*. Bloomington: Indiana University Press.

Haraway, Donna. 1997. *Modest_Witness@Second_Millennium.FemaleMan©_Meets_ OncoMouse™: Feminism and Technoscience*. New York: Routledge.

Huffer, Lynne. 2013. *Are the Lips a Grave?: A Queer Feminist on the Ethics of Sex*. New York: Columbia University Press.

Martin, Emily. 1991. "The Egg and the Sperm: How Science Has Constructed a Romance Based on Stereotypical Male-Female Roles." *Signs: Journal of Women in Culture and Society* 16 (3): 485–501.

McKittrick, Katherine. 2014. *Sylvia Wynter: On Being Human as Praxis*. Durham: Duke University Press.

Oudshoorn, Nelly. 1994. *Beyond the Natural Body: An Archaeology of Sex Hormones*. New York: Routledge.

Richardson, Sarah S. 2012. "Sexing the X: How the X Became the 'Female Chromosome.'" *Signs: Journal of Women in Culture and Society* 37 (4): 909–933.

Somerville, Siobhan B. 2000. *Queering the Color Line: Race and the Invention of Homosexuality in American Culture*. Durham: Duke University Press.

Stoler, Ann Laura. 1995. *Race and the Education of Desire: Foucault's History of Sexuality and the Colonial Order of Things*. Durham: Duke University Press.

Subramaniam, Banu. 2009. "Moored Metamorphoses: A Retrospective Essay on Feminist Science Studies." *Signs: Journal of Women in Culture and Society* 34 (4): 951–980.

———. 2014. *Ghost Stories for Darwin: The Science of Variation and the Politics of Diversity*. Urbana-Champaign: University of Illinois Press.

Weheliye, Alexander G. 2014. *Habeas Viscus: Racializing Assemblages, Biopolitics, and Black Feminist Theories of the Human*. Durham: Duke University Press.

Willey, Angela. 2016. *Undoing Monogamy: The Politics of Science and the Possibilities of Biology*. Durham: Duke University Press.

Sexing the X

How the X Became the "Female Chromosome"

SARAH S. RICHARDSON

"UNEXPECTED." "COUNTERINTUITIVE." "INTELLECTUALLY SURPRISing" (Kuman 2001). These were among the exclamations of researchers upon the 2001 discovery that the human X chromosome carries a large collection of male sperm genes (Wang et al. 2001). Although both males and females possess an X chromosome, the X is frequently typed as the "female chromosome" and researchers assume it carries the genes for femaleness. This essay traces the origins of this long-standing and infrequently questioned association of the X with femaleness and examines the influence of this assumption on historical and contemporary genetic theories of sex and gender difference.

Humans possess twenty-two pairs of autosomal chromosomes and one pair of sex chromosomes—X and Y for males, X and X for females. Today it is well established that the Y carries a critical genetic switch for male sex determination. The X, however, has no parallel relationship to femaleness. Female sexual development is directed by hormones acting in concert with genes carried by many chromosomes and is not localized to the X. Indeed, the X is arguably more important to male biology, given the large number of X-linked diseases to which men are uniquely exposed. Despite this, researchers attribute feminine behavior to the X itself and assume that female genes and traits are located on it. Researchers look to the X to

---

Reprinted and abridged with permission from Sarah S. Richardson, "Sexing the X: How the X Became the 'Female Chromosome,'" *Signs: Journal of Women in Culture and Society* 2012, vol. 37, no. 4: 909–933

explain sex differences and female quirks and weaknesses and have argued that men are superior because they possess one fewer X than females.

The X chromosome offers a poignant example of how the gendering of objects of biological study can shape scientific knowledge. Moving freely between stereotypical conceptions of femininity and models of the X chromosome, X-chromosomal theories of sex differences reveal a circular form of reasoning that is familiar in gender analysis of biology. As Evelyn Fox Keller writes: "A basic form common to many [feminist analyses of science] revolves around the identification of synecdochic (or part for whole) errors of the following sort: (a) the world of human bodies is divided into two kinds, male and female (i.e., by sex); (b) additional (extraphysical) properties are culturally attributed to these bodies (e.g., active/passive, independent/dependent, primary/secondary: read *gender*); and (c) the same properties that have been ascribed to the whole are then attributed to the subcategories of, or processes associated with, these bodies" (1995, 87). A classic historical example of this phenomenon is the gendering of the egg and sperm in mid-twentieth-century medical textbooks, documented by Emily Martin (1991). A second example is the gendering of the sex steroids estrogen and testosterone, as told by Nelly Oudshoorn (1994) and Anne Fausto-Sterling (2000).

Rooted in history and philosophy of science, and drawing on the interdisciplinary methods and questions of feminist science studies forged by scholars such as Fausto-Sterling, Keller, Donna Haraway, and Martin, this essay investigates the sexing of the X in a variety of scientific materials both internal and external to the biosciences. The sexing of the X, I argue, represents a case of gender-ideological bias in science, both historically and in the present day. [. . .]

The Feminine Chromosome

Scientific and popular literature on the sex chromosomes is rich with examples of the gendering of the X and Y. The X is dubbed the "female chromosome," takes the feminine pronoun "she," and has been described as the "big sister" to "her derelict brother that is the Y" (Vallender, Pearson, and Lahn 2005, 343), and as the "sexy" chromosome (Graves, Gecz, and Hameister 2002). The X is frequently associated with the mysteriousness and variability of the feminine, as in a 2005 *Science* article headlined "She

Moves in Mysterious Ways" and beginning, "The human X chromosome is a study in contradictions" (Gunter 2005, 279). The X is also described in traditionally gendered terms as the more sociable, controlling, conservative, monotonous, and motherly of the two sex chromosomes. Similarly, the Y is a "he" and ascribed traditional masculine qualities—macho, active, clever, wily, dominant, as well as degenerate, lazy, and hyperactive.[1]

There are three common gendered tropes in popular and scientific writing on the sex chromosomes. The first is the portrayal of the X and Y as a heterosexual couple with traditionally gendered opposite or complementary roles and behaviors. For instance, MIT geneticist David Page says, "The Y married up, the X married down. . . . The Y wants to maintain himself but doesn't know how. . . . He's falling apart, like the guy who can't manage to get a doctor's appointment or can't clean up the house or apartment unless his wife does it" (Dowd 2005). Biologist and science writer David Bainbridge (2003) describes the evolutionary history of the X and Y as a "sad divorce" (56) set in motion when the "couple first stopped dancing," after which "they almost stopped communicating completely" (58). The X is now an "estranged partner" of the Y, he writes, "having to resort to complex tricks" (145). Oxford University geneticist Brian Sykes (2003) similarly describes the X and Y as having a "once happy marriage" (283–84) full of "intimate exchanges" (42–43) now reduced to only an occasional "kiss on the cheek" (44). A 2006 article on X-X pairing in females in *Science* by Pennsylvania State University geneticist Laura Carrel is headlined "'X'-Rated Chromosomal Rendezvous."

Second, sex chromosome biology is often conceptualized as a war of the sexes. In Matt Ridley's *Genome: The Autobiography of a Species in 23 Chapters* (1999), the chapter on the X and Y chromosomes is titled "Conflict" and relates a story, straight from *Men Are from Mars, Women Are from Venus* (Gray 1992), of two chromosomes locked in antagonism and never able to understand each other (107). A 2007 *ScienceNOW Daily News* article similarly insists on describing a finding about the Z chromosome in male birds (the equivalent of the X in humans) as demonstrating "A Genetic Battle of the Sexes" (Pain 2007), while Bainbridge (2003) describes the lack of a second X in males as a "divisive . . . discrepancy between boys and girls" (83), a genetic basis for the supposed war of the sexes.

Third, sex chromosome researchers promote the X and Y as symbols of maleness and femaleness with which individuals are expected to identify and in which they might take pride. Sykes offers the Y chromosome as a totem of male bonding, urges males to celebrate their unique Y-chromosomes, and calls for them to join together to save the Y from extinction in his 2003 *Adam's Curse: A Future without Men*. Females are also encouraged to identify with their Xs. Natalie Angier (1999) urges that women "must take pride in our X chromosomes. . . . They define femaleness" (26). The "XX Factor" is a widely syndicated column about women's work/life issues on Slate.com, with the slogan "What Women Really Think"; it is also the name of an annual competition for female video gamers (Slate). The promotional video for the Society for Women's Health Research (2008), designed to convince the viewer of how very different men and women really are, is titled "What a Difference an X Makes!"

How the X became the Female Chromosome

The notion of the X as the female chromosome arises from its history as an object of research and its ensuing gendered valence within biological and popular theories of sex. It was originally assumed that the X, not the Y, was the sex-determining chromosome in humans. [. . .]

Historically contingent technical and material factors also helped to brand the X as female. The dominance of studies of the fruit fly *Drosophila* in the first half-century of genetic research played a central role. Unlike in mammals, in *Drosophila* the X is female determining. This is a threshold effect, in which sex is determined by the ratio of autosomes to X chromosomes, with more Xs producing femaleness. In textbook explanations of sex chromosomes from the first quarter of the century, an ink drawing of *Drosophila* chromosomes was ubiquitously used to illustrate the section on the chromosomal theory of sex (Morgan 1915, 7; Wilson 1925). [. . .] The *Drosophila* model suggested that in humans, as in flies, the X should be expected to determine femaleness. In the early days chromosomes were also studied almost exclusively in male gametes—the sperm. Looking at sperm, which as reproductive cells possess only one member of each chromosome set, a perfect dichotomy appeared: half the sperm cells had the X, and half did not. This led to a hyperbinary view of the X and Y. The sperm with an X always produces a female, and the X in the males' sperm is always inherited from the female parent. Failing to distinguish between the "sex" of the gamete and the sex of the organism, this distorted perspective helped to prematurely assign the X to femaleness.

[. . .] The focus on sperm introduced a bias into early sex chromosome research. The centrality of maleness and male tissue to this research led scientists to the conclusion that the X is female and the Y is male. Had researchers looked at somatic tissue, the dichotomy would have been far less clear-cut: both males and females possess at least one X. [. . .]

The first significant breakthrough for human sex chromosome research was the identification of a condensed body present only in female cells. Discovered in 1949, the Barr body, an artifact of the presence of two X chromosomes, suddenly allowed nuclear sexing of any human cell (Barr and Bertram 1949). Murray Barr described the revelation that the "nuclei bear a clear imprint of sex" (Barr 1959, 681) as the "principle of nuclear sexual dimorphism" (682). The notion that every cell has a sex shifted the terms of human sex research and ushered sex difference into the genetic age. [. . .]

By the 1960s, human sex chromosome aneuploidies and other chromosomal anomalies had become potent symbols of the fascinating and exciting new genetics. The historian of midcentury genetics Soraya de Chadarevian (2006, 724–25) argues that this chromosome symbolism, along with the representational schema of the human karyotype, was the public icon of modern genetics in the 1950s and 1960s, before the double helix took its place. It was through this imagery, and the novelty of sex chromosome aneuploidies, that the public first became widely conscious of the X and Y as the molecular pillars of biological femaleness and maleness.

The official findings of human cytogenetics of the 1950s and 1960s were as follows: Human males and females possess twenty-two pairs of autosomes and a pair of sex chromosomes. Males have an X and a Y, and females have two X chromosomes. In females one X in each cell is inactivated early in development, equalizing dosage of X-chromosomal genes in males and females. Subsequent research revealed that the Y chromosome primarily carries a gene that initiates male sexual development and bears few other genes. In contrast, the X chromosome is similar to an autosome, with more than a thousand genes. The X plays no special role in female development, which is controlled by a variety of genes on several different chromosomes.

The idea that the X was female determining was promptly discarded in light of these new findings. The female or feminine resonance that had accumulated around the X chromosome, however, did not fall away. [. . .] Old habits and the force of the idea of a molecular gender binary revealed in the X and Y were irresistible. As the Y would be the male chromosome, the X would continue to be the female one. [. . .]

Tracking the Female X into Human Genetics

The cases of Turner and Klinefelter syndromes demonstrate how the idea of the female-engendering X was carried forward into the human genetics era and how the notion of the female chromosome continued to inflect reasoning about human health and biology even after the X was found not to determine femaleness in humans. Both Turner and Klinefelter were well-documented syndromes of gonadal dysgenesis prior to human chromosome research. Physicians in the United States identified Turner syndrome in 1938 as a syndromic—meaning characterized by a complex of symptoms not localized to any single organ system—phenotype found exclusively in women. Traits included short stature, infertility, and neck webbing (Turner 1938). A Massachusetts General Hospital physician described Klinefelter syndrome in 1942 as a disorder of gonadal underdevelopment in males, resulting in hormonal deficiencies causing infertility and limited body hair (Klinefelter Jr, Reifenstein Jr, and Albright Jr. 1942).

Barr body screening in the 1950s revealed that Turner females lack a second X and that Klinefelter males carry an extra X. Once associated with sex chromosome aneuploidy in the 1950s, the disorders were re-described in more strongly sexed and gendered terms. The infertility of the XO Turner woman was portrayed as evidence of her masculinity rather than a disorder of female sexual development and of development in general. Turner women were claimed to have masculine cognitive traits such as facility with spatiality, discomfort with female gender roles, and defeminized body shape. XXY Klinefelter males were portrayed as feminine, with much emphasis on their purportedly un-muscular body frame, female body fat distribution, lack of body hair, and infertility. The eminent

British geneticist Michael Polanyi even proposed that XO females were "sex reversed males" (Harper 2006, 79). Patricia A. Jacobs and John Anderson Strong (1959) described an XXY individual as "an apparent male... with poor facial hair-growth and a high-pitched voice" (302). They continued, "There are strong grounds, both observational and genetic, for believing that human beings with chromatin-positive nuclei are *genetic females* having two X chromosomes. The possibility cannot be excluded, however, that the additional chromosome is an autosome carrying feminizing genes" (302). A 1967 *New York Times* article similarly captures this mode of reasoning. With the headline "If her chromosomes add up, a woman is sure to be woman," it describes XXY males as having "a few female traits" (Brody 1967, 28). Studies were even undertaken to determine whether Turner women show a tendency toward lesbianism or Klinefelter men incline toward homosexuality or cross-dressing.[2]

These assumptions about the X as feminizing distorted understanding of these disorders, stigmatized individuals carrying them, and misdirected research and clinical care. Today, clinicians specializing in Klinefelter and Turner management emphasize that these are not diseases of gender confusion. Klinefelter patients are phenotypic males, and Klinefelter is not a syndrome of feminization. We now know that Klinefelter is one of the most common genetic abnormalities and often has so few manifestations that men live out their lives never knowing of their extra X. Writes Robert Bock (1997), "For this reason, the term 'Klinefelter syndrome' has fallen out of favor with medical researchers. Most prefer to describe men and boys having the extra chromosome as 'XXY males.'" Similarly, XOs are phenotypic females. Turner syndrome, which has more profound and systemic phenotypic effects than XXY, is emphatically not a masculinizing condition. Physical deformities, heart trouble, infertility, and, occasionally, social and cognitive difficulties are the principal concerns for Turner females.

Throughout the history of twentieth-century genetics, gendered conceptions of the X chromosome fueled ideological conceptions of femaleness and maleness. Today the conception of the X as the female chromosome is not obsolete. It remains a common assumption in twenty-first-century genomics and a source of distortion and bias in genetic reasoning. [...] Perhaps the most prominent case of how the sexing of the X as female continues to operate today, however, is found in "X mosaicism" theories of female biology, health, and behavior. [...]

Gender in X Mosaicism Research

From its inception, the hypothesis that females are cellular mosaics for X chromosomal genes was received as confirmation of dominant cultural assumptions about gender difference. The characterization of females as mosaics or chimeras resonated with conceptions of women as more mysterious, contradictory, complicated, emotional, or changeable.[3] The future Nobel laureate molecular biologist Joshua Lederberg wrote in 1966, "The chimerical nature of woman has been a preoccupation of poets since the dawn of literature. Recent medical research has given unexpected scientific weight to this concept of femininity" (E7).[4] Reporting on the new finding in 1963, *Time* magazine asserted "the cocktail-party bore who laces his chatter with the tiresome cliché about 'crazy, mixed-up women' has more medical science on his side than he knows. . . . Even normal women, it appears, are mixtures of two different types of cells, or what the researchers call 'genetic mosaics'" (1963).

Today, the notion of X mosaicism as scientific confirmation of traditional ideological conceptions of female instability, contradiction, mystery, complexity, and emotionality is thoroughly entrenched. As science writer Nicholas Wade told the *New York Times* in 2005, "Women are mosaics, one could even say chimeras, in the sense that they are made up of two different kinds of cell. Whereas men are pure and uncomplicated, being made up of just a single kind of cell throughout" (Dowd 2005). A 2005 Pennsylvania State University press release similarly announced, "For every man who thinks women are complex, there's new evidence they're correct; at least when it comes to their genes."

These metaphors and gender assumptions are widely shared by present day sex chromosome researchers. Duke University geneticist Huntington Willard, for instance, is quoted saying, "Genetically speaking, if you've met one man, you've met them all. We are, I hate to say it, predictable. You can't say that about women," and Massachusetts Institute of Technology geneticist David Page says, "Women's chromosomes have more complexity, which men view as unpredictability" (Dowd 2005). British geneticist Robin Lovell-Badge has similarly said that "10% [of genes on the X] are sometimes inactivated and sometimes not, giving a mechanism to make women much more genetically variable than men. I always thought they were more interesting!" (Kettlewell 2003).

Barbara Migeon, the Johns Hopkins X chromosome geneticist [. . .] and author of the book *Females Are Mosaics* (2007), is a leading promoter of the theory that X mosaicism is a fundamental mechanism of sex differences and a hallmark of female biology and behavior. Migeon claims "somatic cellular mosaicism . . . has a profound influence on the phenotype of mammalian females" (1994, 230). According to Migeon, X mosaicism "creates biological differences between the sexes that affect every aspect of their lives, not just the sexual ones" (2007, 211). Migeon proposes that "cellular mosaicism . . . is likely to contribute to some of the gender differences in behavior" (2007, 209), including females' response to humor and differences in aggression, emotionality, and educational performance between males and females (2006, 1432–33). Molecular research on X chromosome mosaicism, Migeon argues, offers a promising platform for uncovering sex differences in the brain that studies of brain anatomy have not, thus far, revealed: "Despite dramatically different behavior between the sexes, surprisingly few anatomical differences have been identified," she writes, "[Perhaps] mosaicism for X-linked genes . . . may contribute to some of these sex differences in behavior" (2007, 211).

These speculative scientific conceptions of X mosaicism and femaleness are present in popular discourse around gender differences. Science reporter Natalie Angier, in *Woman: An Intimate Geography* (1999), celebrates female X chromosome mosaicism as a privilege of womanhood and a source of special womanly qualities. "Every daughter," she writes, "is a walking mosaic of clamorous and quiet chromosomes, of fatherly sermons and maternal advice, while every son has but his mother's voice to guide him" (25). She posits what she calls "the mystical X" as a source of "female intuition" and asserts that women "have . . . with the mosaicism of our chromosomes, a potential for considerable brain complexity" (25). Angier imagines a woman's X chromosomes as animating her brain with conflicting voices: "a woman's mind is truly a syncopated pulse of mother and father voices, each speaking through whichever X, maternal or paternal, happens to be active in a given brain cell" (25).

Female X mosaicism is also invoked to bring the authoritative veneer of molecular science to traditional and pejorative views of femininity. Bainbridge's *The X in Sex: How the X Chromosome Controls Our Lives* (2003), for instance, asserts that X chromosome mosaicism confirms that "women are mixed creatures and men are not . . . in a way far deeper" than previously thought (130). Citing the roots of this notion in the Christian vision of Mary as "both virgin and mother" (129), Bainbridge claims that women "represent some intermediate hybrid state" (128), revealed in their "unpredictable, capricious nature" (127). X mosaicism is a "natural reminder of just how deeply ingrained the mixed nature of women actually is" (148), writes Bainbridge. He continues: "So women's bodies truly are mixed-in a very real way. . . . Each woman is one creature and yet two intermingled" (151). [. . .]

Conclusion

Currently, there is a broad popular, scientific, and medical conception of the X chromosome as the mediator of the differences between males and females, as the carrier of female-specific traits, or otherwise as a substrate of femaleness. As this essay has documented, associations between the X and femaleness are the accumulated product of contingent historical and material processes and events, and they are inflected by beliefs rooted in gender ideology. The still very contemporary view that the double X makes females unpredictable, mysterious, chimeric, and conservative, while the single X allows men to learn, evolve, and have bigger brains but also makes them the more risk taking of the two sexes, shows how conceptions of X chromosome structure and function often reflect and support traditional gender stereotypes.

In light of the empirical and conceptual weaknesses of these theories, scientists must work to develop alternative models of the relationship between the X and sex. They must cultivate an active practice of gender criticality, exposing their theories to rigorous examination from all perspectives. [. . .] As this essay has shown, the X chromosome has not only become female identified as an object of biological research, but has, more broadly, become a highly gendered screen upon which cultural theories of sex and gender difference have been projected throughout the twentieth century and up to the present day. The case of how the X became the female chromosome presents a prominent example of how unquestioned gender assumptions can distort and mislead, not only within the biological sciences but more generally in the production of knowledge.

Notes

1 See, e.g., Burgoyne 1998; Angier 1999, 2007; Graves 2000; Bainbridge 2003.

2 See also Miller, 2006 on the deliberations over the true gender of Turner and Klinefelter individuals in the decade after the discovery of the Barr body.

3 In biology, a genetic mosaic is distinct from a genetic chimera. Mosaics carry two different types of cells, whereas chimeras are made up of fused cells of two individuals or species. "Mosaic" and "chimera" are used interchangeably and with the same connotations in the literature on X mosaicism, however, and I follow suit here.

4 Lederberg also notes, however, that the case of XXY males "complicates the myth that chimerism is femininity" (1966, E7).

Works Cited

1963. "Research Makes It Official: Women Are Genetic Mosaics." *Time*, January 4.

2005. "Men and Women: The Differences Are in the Genes." *ScienceDaily*, March 25. www.sciencedaily.com/releases/2005/03/050323124659.htm.

2008. "What a Difference an X Makes." Society for Women's Health Research, Washington DC.

Angier, Natalie. 1999. *Woman: An Intimate Geography*. Boston: Houghton Mifflin Co.

———. 2007. "For Motherly X Chromosome, Gender is Only the Beginning." *The New York Times* 1.

Bainbridge, David. 2003. *The X in Sex: How the X Chromosome Controls Our Lives*. Cambridge, MA: Harvard University Press.

Barr, Murray L. 1959. "Sex Chromatin and Phenotype in Man." *Science* 130 (3377): 679–685.

Barr, Murray L., and Ewart G. Bertram. 1949. "A Morphological Distinction between Neurones of the Male and Female, and the Behaviour of the Nucleolar Satellite during Accelerated Nucleoprotein Synthesis." In *Problems of Birth Defects*, 101–102. Springer.

Bock, Robert. 1997. *Understanding Klinefelter Syndrome: A Guide for XXY Males and Their Families*: US Department of Health and Human Services, Public Health Service, National Institutes of Health, National Institute of Child Health and Human Development.

Brody, Jane. 1967. "If Her Chromosomes Add Up, a Woman Is Sure to Be a Woman." *New York Times*, 16.

Burgoyne, Paul S. 1998. "The Mammalian Y Chromosome: A New Perspective." *Bioessays* 20 (5): 363–366.

Carrel, Laura. 2006. "'X'-Rated Chromosomal Rendezvous." *Science* 311 (5764): 1107–1109.

De Chadarevian, Soraya. 2006. "Mice and the Reactor: The 'Genetics Experiment' in 1950s Britain." *Journal of the History of Biology* 39 (4): 707–735.

Dowd, Maureen. 2005. "X-celling over Men." *The New York Times*, 4–13.
Fausto-Sterling, Anne. 2000. *Sexing the Body: Gender Politics and the Construction of Sexuality*. 1st ed. New York: Basic Books.
Graves, Jennifer A Marshall, Jozef Gecz, and Horst Hameister. 2002. "Evolution of the Human X—A Smart and Sexy Chromosome That Controls Speciation and Development." *Cytogenetic and Genome Research* 99 (1–4): 141–145.
Graves, Jennifer A Marshall. 2000. "Human Y Chromosome, Sex Determination, and Spermatogenesis—A Feminist View." *Biology of Reproduction* 63 (3): 667–676.
Gray, John. 1992. *Men are from Mars, Women are from Venus: A Practical Guide for Improving Communication and Getting What You Want in Your Relationships*. 1st ed. New York: HarperCollins.
Gunter, Chris. 2005. "She Moves in Mysterious Ways." *Nature* 434 (7031): 279–80.
Harper, Peter S. 2006. *First Years of Human Chromosomes: The Beginnings of Humen Cytogenetics*. Bloxham: Scion.
Jacobs, Patricia A, and John Anderson Strong. 1959. "A Case of Human Intersexuality Having a Possible XXY Sexdetermining Mechanism." *Nature* 183: 302–303.
Keller, Evelyn Fox. 1995. "The Origin, History and Politics of the Subject Called 'Gender and Science'" In *Handbook of Science and Technology Studies*, edited by Sheila Jasanoff, Gerald E. Markle, James C. Petersen and Trevor J. Pinch, 80–94.
Kettlewell, Julianna. 2003. "Female Chromosome Has X Factor." *BBC News Online*, March 16. http://news.bbc.co.uk/2/hi/science/nature/4355355.stm.
Klinefelter Jr, Harry F, Edward C Reifenstein Jr, and Fuller Albright Jr. 1942. "Syndrome Characterized by Gynecomastia, Aspermatogenesis without A-Leydigism, and Increased Excretion of Follicle-Stimulating Hormone 1." *The Journal of Clinical Endocrinology & Metabolism* 2 (11): 615–627.
Kuman, Seema. 2001. Genes for Early Sperm Production Found to Reside on X Chromosome. Cambridge, MA: MIT press release.
Lederberg, Joshua. 1966. "Poets Knew It All Along: Science Finally Finds Out That Girls Are Chimerical; You Know, Xn/Xa." *Washington Post*, 18.
Martin, Emily. 1991. "The Egg and the Sperm: How Science Has Constructed a Romance Based on Stereotypical Male-Female Roles." *Signs* 16 (3): 485–501.
Migeon, Barbara R. 1994. "X-Chromosome Inactivation: Molecular Mechanisms and Genetic Consequences." *Trends in Genetics* 10 (7): 230–235.
———. 2006. "The Role of X Inactivation and Cellular Mosaicism in Women's Health and Sex-Specific Diseases." *Jama* 295 (12): 1428–1433.
———. 2007. *Females are Mosaics: X Inactivation and Sex Differences in Disease*. New York: Oxford University Press.
Miller, Fiona Alice. 2006. "'Your True and Proper Gender': The Barr Body as a Good Enough Science of Sex." *Studies in History and Philosophy of Science Part C: Studies in History and Philosophy of Biological and Biomedical Sciences* 37 (3): 459–483.

Morgan, Thomas Hunt. 1915. *The Mechanism of Mendelian Heredity*. New York: Holt.
Oudshoorn, Nelly. 1994. *Beyond the Natural Body: An Archaeology of Sex Hormones*. New York: Routledge.
Pain, Elisabeth. 2007. "A Genetic Battle of the Sexes." *ScienceNOW Daily News*, March 22. http://news.sciencemag.org/sciencenow/2007/03/22-04.html.
Ridley, Matt. 1999. *Genome: The Autobiography of a Species in 23 Chapters*. 1st U.S. ed. New York: HarperCollins.
Slate. "The XX Factor: What Women Really Think." www.slate.com/blogs/xx_factor.html.
Sykes, Bryan. 2003. *Adam's Curse: A Future without Men*. New York: Bantam Press.
Turner, Henry H. 1938. "A Syndrome of Infantilism, Congenital Webbed Neck, and Cubitus Valgus 1." *Endocrinology* 23 (5): 566–574.
Vallender, Eric J, Nathaniel M Pearson, and Bruce T Lahn. 2005. "The X Chromosome: Not Just Her Brother's Keeper." *Nature Genetics* 37 (4): 343–345.
Wang, P Jeremy, John R McCarrey, Fang Yang, and David C Page. 2001. "An Abundance of X-Linked Genes Expressed in Spermatogonia." *Nature Genetics* 27 (4): 422–426.
Wilson, Edmund B. 1925. *The Cell in Development and Heredity*. New York: Macmillan.

Pelvic Politics

Sexual Dimorphism and Racial Difference

SALLY MARKOWITZ

WHAT, EXACTLY, DOES THE INTERSECTION OF RACE AND SEX/GENDER ideologies come to? Can we describe this intersection in a systematic way? These are not easy questions. At the very least, historians tell us that "scientific" classifications of race and sex have long been associated with each other: in temperament, intelligence, and physiology, so-called lower races have often provided a metaphor for the female type of humankind and females a metaphor for the "lower race" of gender (Stepan 1993). But, I argue, this analysis does not give the whole story either. While "lower" races may often be represented as feminine and the men of these races as less than masculine, the femininity of nonwhite women, far from being heightened, is likely to be denied (the better, no doubt, to justify their hard physical labor or sexual exploitation). Indeed, to talk simply about metaphorical connections between the discourses of race and sex may even be to overstate the autonomy of each. It is not difficult, after all, to find a pronounced racial component to the idea(l) of femininity itself: to be truly feminine is, in many ways, to be white. And if a woman is not white, her oppression in such crucial areas as reproductive life, sexuality, and work cannot be separated easily into discrete racial and gender components without distorting its character (Davis 1981; Carby 1987; Hill Collins 1990; hooks 1990). In light of such phenomena, the question is

Reprinted and abridged with permission from Sally Markowitz, "Pelvic politics: Sexual dimorphism and racial difference," *Signs: Journal of Women in Culture and Society* 2001, vol. 26, no. 2: 389–414

bound to arise: Might we have missed some more systematic, underlying connection between the ideologies of race and sex/gender?

[. . .] Whatever their other connections, hierarchical social classifications based on race and those based on sex/gender have long been connected through the category of sex/gender *difference*. That is to say, in dominant Western ideology a strong sex/gender dimorphism often serves as a human ideal against which different races may be measured and all but white Europeans found wanting. This ideal then functions as a measure of a racial advancement that admits of degrees determined by the (alleged) character of the relationship between men and women *within a particular race*. And so, I argue, phenomena such as femininity's "whiteness" in dominant ideology (and, alas, in feminism) or the inextricability of the racial and sex/gender oppression in some women's lives turn out to be elements of a larger ideological structure in which sex/gender difference is imagined to increase as various races "advance." If this is so, moreover, the ideology of sex/gender difference itself turns out to rest not on a simple binary opposition between male and female but rather on a scale of racially coded degrees of sex/gender difference culminating in the manly European man and the feminine European woman. Thus, *even in ideology*, sex/gender difference is not as binary as it might at first seem, and so the feminist project of displacing the sex/gender binary—of "thinking beyond sexual dimorphism" (to borrow from the title of a recent anthology, Herdt 1994) by destabilizing such dimorphism or rejecting it outright—misses a very important point: An ideology that considers sexual dimorphism to be embodied only in European "races" has *already*, in a sense, thought beyond it—hardly, it starts to seem, a revolutionary accomplishment. [. . .]

The racialized notion of sex/gender difference that I explore [. . .] can already be detected, I suspect, in the eighteenth-century emergence of modern conceptions of race and sex/gender themselves.[1] By the late nineteenth century, such racialization had become explicit and widespread. Darwin himself seriously considered it, citing the German race theorist Karl Vogt, who embraced it outright.[2] Indeed, such a view seemed so reasonable that in 1886 the influential sexologist Richard von Krafft-Ebing could simply, without preamble or explanation, say, "The secondary sexual characteristics differentiate the two sexes; they present the specific male and female types. The higher the anthropological development of the race, the stronger these contrasts between man and woman" (28). And what could make more sense? If the display of either a pronounced male or female character is the ideal to which each human is expected to conform, then it stands to reason that the men and women of the most "advanced" race(s) will meet this ideal best. That is to say, just as personhood in our social world is so thoroughly gendered that one must register unambiguously as either a man or a woman (if preferably the former) in order to count as fully human, so too must a *race* display a pronounced sex/gender dimorphism in order to qualify as "advanced."

It would, of course, be surprising to hear this view stated so baldly today, but this hardly means that such racialization has finally withered away, leaving behind an innocuous category of sex/gender dimorphism purified of racial connotations. More likely, this racialization has achieved that familiar if peculiar kind of invisibility with which contemporary liberal ideology veils race and racism themselves. [. . .]

However, the racialization of sex/gender difference has not always been so difficult to discern. Indeed, I take as my point of departure a particularly explicit and revealing expression of it by turn-of-the-century British sexologist Havelock Ellis. Perhaps second only to Freud's in shaping modern conceptions of sexuality and gender, Ellis's work has also, according to Sander Gilman (1985), been central to the modern European sexualization of African women. But even Gilman, it is worth pointing out, overlooks Ellis's more basic racialization of sex *difference*, which fairly leaps from the page once one looks for it. Indeed, that is one reason Ellis warrants attention: rather than simply assuming a racialized sex/gender dimorphism (as Krafft-Ebing, e.g., does), he develops the notion at some length, relying on it to formulate and resolve a set of paradoxes that arise from subjecting a single social world to the competing yet (one cannot help but sense) somehow complementary classifications of race and sex/gender. In particular, Ellis, late Victorian that he was, struggled to maintain a cross-racial sex classification—that is, one that divides the human world exhaustively into men and women—that did not challenge what he believed to be the hierarchical differences between the races. But how, exactly, could one square the racial superiority of European women with both their inferiority to their mates and their similarity to non-European women? And what could an English gentleman be said to share with an African or Asian male? In short, how could the distinction between the sexes both specify and justify bourgeois gender relations in particular even while applying universally across race, class, and culture (Poovey 1988)? Such questions should interest feminists, and not only for the light they shed on the conundrums that follow from a simultaneous commitment to racism and sexism. For the logic behind such questions also shapes some of the most troubling and persistent problems plaguing contemporary feminism, which must, after all, reconcile a theory and politics based on the category of sex/gender with the profound differences among women that race makes.[3] Ellis solved his version of these problems by racializing sex/gender difference itself, a solution that feminists can hardly borrow; indeed, such racialization is obviously part of the problem. But it is a crucial part, and one that feminists risk perpetuating by overlooking.

It is not surprising that Ellis develops his views on race and sex/gender in the context of an analysis of feminine beauty, another physiological discourse heavily freighted with ideological significance. Indeed, his analysis draws together a number of important, disparate strands—the tensions between female sexuality, domesticity, and maternity; the relationship between sex and gender; the role of race in the discourse of female beauty; anxiety about "interracial mixing"—all of which seem to converge in Ellis's *fin-de-siècle* contribution to a confusing just-so story of the female pelvis that dates back to the eighteenth-century quest for the pelvic marks of racial identity and hierarchy. Not surprisingly, this story was equivocal from the start, offering various and conflicting views about what exactly constituted the racially advanced pelvis. On the one hand, some early anthropologists and physiologists claimed that the wide female pelvis (so prized later by Ellis) signified racial "primitivism," since a generous pelvis seemed to promote the ease in childbirth supposedly enjoyed by beasts—a convenient justification for continuing to drive hard-laboring female slaves of "lower race" even when they were pregnant (Schiebinger 1993, 156–58). On the other hand, by 1826 Willem Vrolik had linked a wide pelvis with racial superiority, a view appropriated later in the century by Ellis and his ilk, who, ingeniously combining Darwinism with craniometry, insisted that as races became more advanced, their increased head size required a wider maternal pelvis to accommodate the larger skull of the racially superior infant (Gilman 1985, 90; Schiebinger 1993, 156–59).

Indeed, for Ellis and his contemporaries, the generous pelvis also promised all the delights of normative femininity: maternal fitness, gentleness, domesticity, beauty. So although I have been using the term *sex/gender difference,* it is worth emphasizing that Ellis, like his heirs the sociobiologists, regards what we have learned to call gender (a psychosocial category) as rooted firmly in sex (a physiological one). Thus, what is racialized for Ellis and others is not just *gender* difference, the various degrees and manifestations of which might reasonably be thought to depend on culture, but actual *sex* difference as well. This view may seem quite startling today since it appears to fly in the face of the contemporary doctrine of sexual dimorphism, which posits the fundamental, exhaustive, and compelling physiological distinction between the male and the female sex, each with its own thoroughgoing physiological essence. Certified by scientists and accepted without question by nearly everyone else, this doctrine is usually taken to be an all-or-nothing affair, applying universally (and hence, in a sense, democratically) across races. But we should be cautious about accepting at face value the apparent race-neutrality of such conceptions of physiological sex. Indeed, recent work by feminists and others suggests that, far from determining gender, some of the supposedly objective physiological facts of human sex may look as they do only when viewed, described, and organized through gender's lens (Kessler and McKenna 1978; Fausto-Sterling 1985; Butler 1990; Garber 1992; Herdt 1994). And so if we acknowledge how thoroughly the ideology of race continues to saturate that of gender, it would be surprising if the "scientific" category of sexual dimorphism altogether escaped gender ideology's long arm even today.

In any case, we should remember that for all of its appearance of empirical fact, the doctrine of sexual dimorphism is also a historically situated ideology; indeed, it is a fairly recent one at that, as Thomas Laqueur shows (1990). Laqueur acknowledges that gender dualism—the system of social and political distinctions between men and women—may be ancient, but, he argues, until the eighteenth century this dualism of gender was not usually thought to rest primarily on a dualism of physiological sex. Instead, only one physiological sex was recognized—the male—while the female body was regarded as an inferior version of it, the vagina simply an inverted penis, and menstruation and lactation physiological processes that would, in a body with sufficient "heat" produce not blood or milk but sperm—that precious, rarified form of the same stuff. Not until the mid-eighteenth century does femininity emerge as a full-fledged essence, complementary to masculinity and seemingly inseparable from women's special physiology (Laqueur 1990). This two-sex model, of course, has been central to gender ideology since the eighteenth century; my analysis of Ellis will show its complex contribution to racial ideology as well. [. . .]

Sex and beauty may be intertwined, Ellis tells us, but a civilized appreciation of beauty requires acknowledging—as all but savages in a "low level of culture" are able to do—that the genitals are simply ugly (Ellis 1905, 1:161). Because of their function, sexual organs must retain their "primitive" character; unable to "be greatly modified by sexual or natural selection," they can hardly "be regarded as beautiful" (1:169). That which is beautiful, again, is that which evolves through sexual selection. (Notice too the artistry attributed to generations and generations of European males, who, through sexual selection after sexual selection, actually *create* the master-piece of European female beauty.) In explaining the ugliness of genital form as a consequence of its subservience to function, Ellis seems to echo the Kantian aesthetic ideology that measures the beautiful by its distance from the useful or (re)productive. Nonetheless, feminine beauty, far from autonomous, is inextricably linked for Ellis with maternity: women's beauty turns out to depend not on primary (i.e., genital) sexual characteristics but secondary ones—wide pelvis and full breasts, designed for bearing and nursing. Here Ellis may seem to be moving toward a universal, cross-racial sexual dimorphism, since hips and breasts are, one might assume, universal female endowments. But, according to Ellis, while the *ideal* of full breasts and pelvis may be recognized nearly universally, it is best *realized* by white European women—perhaps because sexual selection, at least as far as the broad pelvis is concerned, coincides with natural selection (1:156). "Broad hips, which involve a large pelvis, are necessarily," he claims, "a characteristic of the highest human races, because the races with the largest heads must be endowed also with the largest pelvis to enable their large heads to enter the world" (1:165). Thus, the broad European pelvis, beautiful and desirable on its own, gains moral dignity through its association with the European (male?) infant's large brain.

This full-breasted, wide-pelvised standard of beauty is an objective one, Ellis claims, accepted even among races in which women lack these features. Indeed, African women, whose pelvis is "the least developed, the narrowest, and the flattest," cultivate steatopygia, "a simulation of the large pelvis of the higher races" consisting of "an enormously exaggerated development of the subcutaneous layer of fat which normally covers the buttocks and upper parts of the thighs in woman" (1:165). True beauty, the female physiological expression of racial superiority, can be only mimicked by less highly evolved races, much as a deceitful prostitute (herself associated with steatopygia in the nineteenth century) might use cosmetics or clothing to disguise the inevitable signs of her depravity (Gilman 1985, 76–108).

But a universal standard of beauty is not the whole story for Ellis. There is also a tendency for "the specific characters of the race or nation" to cause divergence in ideals of beauty, which is "often held to consist in the extreme development of these racial or national anthropological features" (Ellis 1905, 1:210): "It frequently happens that this admiration for racial characteristics leads to the idealization of features which are far removed from aesthetic beauty. The firm and rounded breast is certainly a feature of beauty, but among many of the black peoples of Africa the breasts fall at a very early period, and here we sometimes find that the hanging breast is admired as beautiful" (1:176). So, within any race, universally objective standards of beauty coexist with racially particular ones, and while the more advanced men of lower races may be drawn to the European type, they will be in the minority.

Here, then, is a particular formulation of a tension that continues to be familiar to liberal humanism: that between human universal and human particular, the universal connoting the highest human values, embodied, as always, in the European body and mind, and the particular connoting those partial and parochial values that shape the bodies and minds of the rest of the species. This opposition between universal and particular, between essence and accident—as usual, a problem only for the lower races—is resolved by the civilized white European, in whose mind and body the universal ideal and the existing particular converge.[4]

But this resolution raises new problems. While the match between the European woman's wide pelvis and the European man's appreciation of it establishes a kind of racial reproductive closure based on both sexual and natural selection, Ellis has some difficulty reconciling this feminine ideal with a more obvious indication of "superior" race: skin and hair color. On the one hand, the two seem to go together. The fair woman, claims

Ellis, is universally and rightly agreed to be the most beautiful. Not only does her "brilliantly conspicuous" golden hair complement the "soft outlines of woman," but she is also most likely, on his account, to exhibit that most beautiful and desirable combination of well-developed breasts and wide pelvis (presumably because of her racial superiority) (1:210).

But blondness also has another, differently gendered meaning for Ellis, who complicates his discussion by drawing on a human aesthetic ideal, articulated frequently since the eighteenth century, that associates human beauty with an intellectual and moral perfection that is decidedly *masculine.* Indeed, in the midst of a discussion devoted almost entirely to feminine beauty, he asserts that the male body is actually aesthetically superior to the female, apart from the unfortunate if decisive fact that the protuberance of the male genital organ, especially when erect, ruins the male form—a failing avoided by Woman, whose "sexual region is almost imperceptible in any ordinary and normal position of the nude body" (1:162). So the virtues of beauty, masculinity, and racial superiority are combined, indeed conflated, in the European male, and the one-sex model of humanity turns out to have a race as well (Gobineau 1915; Stepan 1993; Young 1995; 99–117).[5] [. . .]

[For Ellis,] men must be masculine, to be sure, but women, for their part, must be feminine—as indeed they are among the "more refined" races of Britain, where the sexual dimorphism of the fair saves the day. Unlike other marks of femininity—lack of intellectual power and drive, for example—wide pelvises and full breasts are not found in men of the lower races and so cannot be associated simply with racial inferiority. Nor, for that matter, are they found in women of lower races. Instead, they signify a racial superiority of a distinctly feminine sort, one that complements (without, of course, quite equaling) masculine racial superiority. Thus Ellis shows how the move from a one-sex to a two-sex model of humanity is implicated not only in an essentialist notion of woman's physiological difference but also in a theory of racial supremacy that is based on sex/gender dimorphism. [. . .]

This understanding of racialized sexual dimorphism continued to serve racist ideology well into the twentieth century. In 1920, for example, the Viennese anthropologist Robert Stigler remarked on the vagueness of sexual characteristics in Jews, among whom "the women are often found to have a relatively narrow pelvis and relatively broad shoulders and the men to have broad hips and narrow shoulders." Moving without hesitation from the physiological to the social, he notes further that in their advocacy of the "social and professional equality of man and woman," Jews have tried to eliminate the "role secondary sexual characteristics instinctively play among normal people" (Gilman 1993, 162–63).[6] As the century progressed, however, the racialization of sex/gender dimorphism became more subtle. At midcentury anthropologists were still writing about the pelvis, which, no longer the explicit focus of an overtly racist classificatory scheme, now appeared to signify sex/gender difference alone. By 1957 the physical anthropologist Lucile Hoyme could marvel that anthropology had taken so long to recognize the pelvis as a universal indicator of sexual difference and speculate that this delay was the consequence of an earlier era's misguided determination to use the pelvis primarily to differentiate among races. But the role of the pelvis in the story of race was not over yet. Having just hailed anthropologists' realization of the pelvis's relevance to sex rather than race, Hoyme suggests, almost in passing, a possible direction for future study: collecting data measuring the comparative disparity in pelvic measurements between men and women *within particular races.* But this, of course, is to smuggle race in through the back door, as the quest for the racial pelvis becomes masked by the subtler, more complicated, and apparently innocent quest for the pelvic measure of sexual dimorphism within particular races.

Of course, such subtlety may be just what is required today, when ignoring race, as Toni Morrison (1992, 10) has commented, is often understood to be a graceful, even generous, liberal gesture. But this reticence shows only that race is embarrassing these days in ways that sex/gender is not. In this ideological climate, what better harbor for a covert racism than the apparently innocent notion of sexual dimorphism, with its imprimatur of scientific neutrality and its unavowed racist lineage? [. . .] In light of the deeply entrenched and persistent racialization of sex/gender dimorphism, what are we to make of feminism's sometimes uncritical acceptance of an abstract notion of sex/gender difference as its basic category of analysis, let alone as a cause for celebration? [. . .]

But there is a further problem with sexual difference, at least as construed by some contemporary theorists. Feminists may continue to argue over whether the master's tools can dismantle the master's house, but some of these tools seem ill suited even to take its measure. Following a variety of post-Hegelian critical perspectives—particularly structuralism, psychoanalysis, and deconstruction—much academic feminist theory accepts without question a metaphysics and methodology based on binary opposition. At first glance, this emphasis may seem fortuitous, meshing well with what appears to be the sharply and irreducibly binary character of sex/gender ideology. Indeed, some regard this binary character as so deeply rooted—in the psyche, in culture, in language—that one can hope at most for its destabilization from within, whether by a utopian reaffirmation of feminine difference [. . .] or by the parodic, reiterative performances of gender that some other feminists celebrate (Butler 1990; Garber 1992). Still other feminists advocate thinking "beyond sexual dimorphism" by calling attention to the secrecy surrounding newborns of indeterminate sex, who must be altered quickly and decisively lest nature itself seem to have produced counterexamples to the very sex-dimorphic order our culture insists is natural (Kessler and McKenna 1978). Or they point to non-Western conceptions of sex, gender, and sexuality that neither necessarily presume nor enforce the existence of a dimorphic, exhaustive system of sex/gender classification (Herdt 1994).[7]

But all of these approaches risk oversimplifying sex/gender ideology by overstating and misconstruing its binary character, which, in fact, seems all the sharper the more its racial dimension is obscured. For although dualism is certainly a central and abiding organizing category in the West, we should not assume that it is the only one. Perhaps just as ubiquitous and persistent is another, stretching from Plato to Darwin and beyond, that understands the world as a great chain of being, brimming over with every manner of thing, continuous, without gaps, and above all hierarchical (Lovejoy 1936). This gradualist metaphysics of plenitude has helped shape modern Western ideologies of social hierarchy, most obviously in the case of race (Jordan 1968, 217–28; Young 1995, 6–19) but also, if more subtly, in the case of sex and gender. It underlies what I have called (following Laqueur) the pre-Enlightenment one-sex understanding of sex/gender according to which (European) women, like the "lower races," were considered to be (European) man's inferiors rather than his complement. Indeed, one might even say that until the notion of feminine complementarity became ascendant, (European) women and men of "lower race" were regarded as similarly inferior human specimens. The introduction of a more robustly dualist gender ideology, as we have seen, complicated and, to some extent, displaced this ideology of feminine inferiority by introducing the idea(l) of feminine complementarity—an ideal that is essential to the structural connection between sex/gender difference and racial hierarchy that I have traced. But, paradoxically, this structural connection itself depends, in a sense, on a new differentiation between the discourses of race and sex/gender. Whereas, previously, women and non-Western men were thought to share a similar inferiority to European man, now women of privileged race, thanks to their "difference," have become man's complement. So while particular races continued to fall somewhere along the hierarchy defined by the great chain of being, femininity in the abstract became more difficult to place, in part because it was no longer understood simply as inferiority and in part because it was not understood, really, to be a property of *all* females after all (Schiebinger 1993, 143–83). Yet, it would be a mistake to understand the advent of feminine complementarity as simply an instance of dualism trumping gradualism as sex/gender ideology's central organizing category. The gradualism that had informed the pre-eighteenth-century one-sex notion of female inferiority still had a role to play, reemerging in the graduated scale of racialized sex/gender difference along which various races could be situated. Racial ideology, then, incorporated the new conception of femininity as one half of a racialized dimorphic sex/gender ideal, even as sex/gender dimorphism took on its full conceptual and ideological meaning only as an ideal that most races fail to meet. Thus, "thinking beyond sexual dimorphism" is far from a nearly impossible, utopian feat, requiring studies of non-Western cultures or physical anomalies even to give us the idea; on the contrary, to grasp a *racialized* binary of sex/gender is already, in a sense, to have thought beyond what now appears to be sex/gender ideology's rather superficial binarism.

Notes

1. I cannot argue for this view here, but it is suggested by what early anthropologists of race have to say about sexual difference among "primitive" races, as well as by the less-than-universal character of the new, eighteenth-century notion of femininity, which applied only to bourgeois European women (see the selections in Eze 1997; as well as the discussion in Schiebinger 1993, 115–83).
2. Laqueur (1990, 208) claims that Darwin quotes Vogt "approvingly" but see Darwin [1871] 1981, 2: 329–30.

3 The most systematic conceptual treatment I know of these problems remains Elizabeth Spelman's *Inessential Woman* (1988). Among the many women of color who have addressed this issue, bell hooks (1981, 1984, 1990) has been particularly influential.

4 Indeed, feminists who are insufficiently attentive to race and other sorts of social differences between women tend to replicate this tension: the white, middle-class, Western, heterosexual, Christian woman is taken to be the paradigm of what a woman should be; thus, she is also the only sort of woman likely to come even close to embodying this paradigm.

5 For discussions of tile role of aesthetics in racialist theories, see Schiebinger 1993, 126–34; Young 1995, 96–97.

6 In 1983, feminist historian Gisela Bock noted that, according to Nazi ideologues, "the difference and polarity between the sexes (reason/emotion, activity/passivity, paid work/housework) is fully developed only in the 'superior' the 'nordic' races; among 'inferior races' including those of low 'hereditary value' the sexes are less differentiated—and thus heavy and cheap labor is good for both" (417).

7 In a sense, such accounts recuperate and rework the very racialization of sex/gender dimorphism that I have been considering: Whereas the racialist ideology to which Ellis subscribes takes the allegedly insufficient sex/gender dimorphism of non-Europeans as a detect, this so-called failure, reconceived as a cultural rather than a physiological difference, may be reinterpreted positively as a less oppressive alternative to the Western ideology of sex/gender dimorphism, which is then revealed to be an unfortunate cultural construction (Herdt 1994, esp. 21–81).

Works Cited

Bock, Gisela. 1983. "Racism and Sexism in Nazi Germany: Motherhood, Compulsory Sterilization, and the State." *Signs* 8 (3): 400–421.

Butler, Judith. 1990. *Gender Trouble: Feminism and the Subversion of Identity*: Psychology Press.

Carby, Hazel V. 1987. *Reconstructing Womanhood: The Emergence of the Afro-American Woman Novelist*. New York: Oxford University Press.

Darwin, Charles. [1871] 1981. *The Descent of Man and Selection in Relation to Sex*. Princeton, NJ: Princeton University Press.

Davis, Angela Y. 1981. *Women, Race, & Class*. 1st ed. New York: Random House.

Ellis, Havelock. 1905. *Studies in the Psychology of Sex*. 2 vols. New York: Random House.

Eze, Emmanuel Chukwudi. 1997. *Race and the Enlightenment: A Reader*. Cambridge, MA: Blackwell.

Fausto-Sterling, Anne. 1985. *Myths of Gender: Biological Theories about Women and Men*. New York: Basic Books.

Garber, Marjorie B. 1992. *Vested Interests: Cross-Dressing & Cultural Anxiety*. New York: Routledge.
Gilman, Sander L. 1985. *Difference and Pathology: Stereotypes of Sexuality, Race, and Madness*. Ithaca, NY: Cornell University Press.
———. 1993. *Freud, Race, and Gender*. Princeton, NJ: Princeton University Press.
Gobineau, Arthur. 1915. *The Inequality of Human Races*. Translated by Adrian Collins. New York: Putnam.
Herdt, Gilbert H., ed. 1994. *Third Sex, Third Gender: Beyond Sexual Dimorphism in Culture and History*. New York: Zone Books.
Hill Collins, Patricia. 1990. *Black Feminist Thought: Knowledge, Consciousness, and the Politics of Empowerment*. Boston: Unwin Hyman.
hooks, bell. 1981. *Ain't I a Woman: Black Women and Feminism*. Boston: South End Press.
———. 1984. *Feminist Theory from Margin to Center*. Boston: South End Press.
———. 1990. *Yearning: Race, Gender, and Cultural Politics*. Boston: South End Press.
Jordan, Winthrop D. 1968. *White over Black: American Attitudes toward the Negro, 1550–1812*. Chapel Hill: University of North Carolina Press.
Kessler, Suzanne J., and Wendy McKenna. 1978. *Gender: An Ethnomethodological Approach*. Chicago: University of Chicago Press.
Krafft-Ebing, Richard von. [1886] 1965. *Psychopathia Sexualis, with Especial Reference to the Antipathetic Sexual Instinct*. Translated by Franklin S. Klaf. New York: Stein and Day.
Laqueur, Thomas Walter. 1990. *Making Sex: Body and Gender from the Greeks to Freud*. Cambridge, MA: Harvard University Press.
Lovejoy, Arthur O. 1936. *The Great Chain of Being: A Study of the History of an Idea*. Cambridge, MA: Harvard University Press.
Morrison, Toni. 1992. *Playing in the Dark: Whiteness and the Literary Imagination, The William E Massey, Sr Lectures in the History of American Civilization*. Cambridge, MA: Harvard University Press.
Poovey, Mary. 1988. *Uneven Developments: The Ideological Work of Gender in Mid-Victorian England*. Chicago: University of Chicago Press.
Schiebinger, Londa L. 1993. *Nature's Body: Gender in the Making of Modern Science*. Boston: Beacon Press.
Spelman, Elizabeth V. 1988. *Inessential Woman: Problems of Exclusion in Feminist Thought*. Boston: Beacon Press.
Stepan, Nancy. 1993. "Race and Gender: The Role of Analogy in Science." In *The "Racial" Economy of Science: Toward a Democratic Future*, edited by Sandra G. Harding, 359–76. Bloomington: Indiana University Press.
Young, Robert. 1995. *Colonial Desire: Hybridity in Theory, Culture, and Race*. New York: Routledge.

The Sexual Reproduction of "Race"

Bisexuality, History, and Racialization

MERL STORR

MY ARGUMENT IN THIS CHAPTER IS THAT THE FORECLOSURE OF THE history of bisexuality must be resisted—not just because bisexuality is an important topic in its own right, [. . .] but also because critical enquiry into bisexuality has particular importance for understanding the constitutive role of "race" in the history of "western" sexuality. Returning to the last *fin de siècle*, I shall focus on two major sexologists: Havelock Ellis, "the most influential of the late Victorian pioneers of sexual frankness" (Rowbotham and Weeks 1977, 141) whose works include the key and, in its day, highly controversial study *Sexual Inversion*; and Richard von Krafft-Ebing, more conservative author of the "canonical" (Smith-Rosenberg 1991, 269) legal-medical textbook *Psychopathia Sexualis*, which was first published in German in 1886 and ran to twelve editions, the last of which appeared in 1903 and will be the focus of my discussion in what follows.

[. . .] Far from being regulated by the dourness of a single binary opposition, the texts of Hirschfeld, Krafft-Ebing and Ellis (and indeed Freud, who was in so many ways their son and heir[1]) are a veritable cornucopia of polymorphous perversities and perversions, marvelously rich and strange in the face of their own attempts to submit their unruly subjects to the rigours of scientific taxonomy. If the hetero/homosexual binary is truly now *the* dominant organizing principle of sexuality, the question of how it emerged into its dominant position from the richness of the nineteenth century is one with which historians of sexuality have yet to get to grips (Cf. Sedgwick 1990, 9). [. . .] The fact that some or even most[2] sexual articulations can now be (and usually are) subsumed under the terms of that binary is, of course, a product of the complex play of power and resistance, both historically and in the present. But as, precisely, an outcome of social, cultural and historical forces, that fact is *contingent*: it is no more necessary, in itself, than the potential subsumption of sexual practices under other organizing principles—vanilla versus SM, say, or polyamorous versus monogamous, or even principles which are not binary at all, such as the focusing of pleasure in particular body parts, or in particular physical environments (cottages, backrooms, sex parties, suburban semis), or in particular modes or states of dress. The production of sexuality, always discursive and localized, is always also multiplicitous and complex, and its reduction to a single binary opposition is a phenomenon that historians and theorists of sexuality should be concerned to interrogate. [. . .]

Thus, for the queer reader of the present *fin de siècle*, the sexological texts of the previous *fin de siècle* appear to offer tantalizing narratives of our polymorphous antecedents; not just homosexuals and bisexuals, but SM practitioners, fetishists, transgendered subjects and many others can find images of their forebears there—speaking to us, as it were, at and from the very moment of their production and regulation by the discourses of science and law. Of course, the search for sexological ancestors must proceed with caution. First, there must be caution against conflating polymorphism with freedom and hence imagining sexuality before the dominance of the hetero/homo binary as a pre-lapsarian flowering of eroticism, when in fact, as a reading of the texts themselves makes plain, the forces of discursive power are very much in play: the very classification of these figures is as much an act of routinizing and regulating desire as is their apparent elision by the hetero/homo binary. Second, there must be caution against the projection of contemporary sexual maps on to the landscapes of the past, expressed at its crassest in the search for sexual identities as if they were historical constants, as if the *Urnings* and *Uranodionings* of the late nineteenth century simply "were" homosexuals and bisexuals in the sense in which we understand those terms today; indeed, the term "bi-sexual" itself then meant something different than it does for us now, as I shall discuss shortly. [. . .]

Historians have already begun to work on uncovering the history of sexuality as marked by the racial positionings of sexual subjects—in particular, the feminist literature on the history of eugenics and of "maternalist imperialism" in Europe and North America is now very extensive (e.g. Bland 1995; Burton 1990; Davin 1978; Ware 1992). In contrast with advances made in this regard, it is arguable that inadequate attention has thus far been paid to the ways in which "race" operates discursively in the production of sexuality *as such*, particularly in the sexological canon which played, and arguably continues to play, a crucial role in the formation of sexuality as we know it today.[3] This neglect of "race" in the sexual discourses of the past reflects and is reflected by the state of certain contemporary sexual-dissident theories, notably queer theory, which, despite the reiteration of "difference" as a key concern, nevertheless fails to encompass racial dynamics at the heart of sexuality. [. . .]

The notion of "race" is a slippery one, in the writings of the sexologists and elsewhere. As Lucy Bland notes of early twentieth-century British usages of the term, "'Race' might mean the 'human race', the 'Anglo-Saxon race', 'the British race', etc., depending on the context. The ambiguity of meaning allowed a slippage between the different usages" (1995, 231). In fact the movements of slippage are complex, with an array of terms unstably distinguished from each other. For example, Krafft-Ebing uses the terms "species" and "race," but the latter has an unstable relationship with the former. Thus he claims that "desires arise in the consciousness of the individual, which have for their purpose the perpetuation of the *species* (sexual instinct). . . . The duration of the physiological processes in the sexual organs, as well as the strength of the sexual desire manifested, vary, both in individuals and in *races*" (Krafft-Ebing [1886] 1965, 16, my italics)—where "race" operates as a subdivision of "species," the purpose of sexual instinct being the perpetuation of the latter; but later in the same text he writes of "the sexual instinct which is indispensable for the preservation of the *race*," (46) with the implication that "race" refers to the single human "race" or "species" rather than to its supposed subdivisions. Havelock Ellis, by contrast, uses "species" relatively rarely, and is more likely to use "race" interchangeably in both senses: "that propagation of the race which, as a matter of fact, we find dominant throughout the whole of life" (Ellis 1897, 130); "we seem to find a special proclivity to homosexuality . . . among certain races and in certain regions" (22).[4]

As I hope these quotations make clear, "race," with all its instability, is the organizing principle of sexuality for these sexologists, procreation and the propagation of "the race" (whatever that might mean at any given point) being the meaning and purpose of sexuality as such. Indeed, the very distinction between the normal and the "perverse" arises from the racial imperative of sex as procreation; according to Krafft-Ebing, "With opportunity for the natural satisfaction of the sexual instinct, every expression of it that does not correspond with the purpose of nature—i.e., propagation—must be regarded as perverse" (52–3). In this the sexologists of the *fin de siècle* are exemplars of nineteenth-century racialism; as Robert Young suggests, "Race became *the* fundamental determinant of human culture and history: indeed, it is arguable that race became the common principle of academic knowledge in the nineteenth century" (Young 1995, 93). It is arguable that all of the sexologists' most important claims arise from this common principle. [. . .]

This is particularly true of bisexuality, the formulation of which in fin-de-siècle sexology can be fully understood only in the context of racialization, and the investigation of which thus highlights the constitutive role of "race" in sexology's understanding of both "normal" and "inverted" sexuality. While there are a number of figures in sexological texts which might have an appeal to contemporary bisexual readers as ancestors—notably the "psychosexual hermaphrodite" attracted to members of both sexes, and the "acquired invert" who, in Krafft-Ebing in particular, appears as a malleable figure who can be persuaded from heterosexuality to homosexuality and, with encouragement from suitable professionals, might be coaxed back again—it is on the term "bi-sexual"[5] itself that I wish to focus here. "Bi-sexual," in both Ellis and Krafft-Ebing, stands in a particular relation to both "race" and the "heterosexual matrix" (Butler 1990), and reveals the mutually constitutive congruence of the two.

"Bi-sexual" in these texts indicates a coincidence of male and female characteristics, and it is primarily a physical phenomenon although it has a number of important psychosexual effects. It is a term which has vital explanatory value for the understanding of the perverse. [. . .]

Bisexuality as a physical phenomenon is not the same as hermaphroditism: the latter is an anomaly of sexual *development*, but the former is the originary state *from* which later developments are made. Krafft-Ebing gives a similar account of bisexuality as an originary state. In a move which echoes the recurrent nineteenth-century claim that "ontogeny recapitulates phylogeny" Krafft-Ebing explicitly locates bisexuality as the origin not just of the development of the individual but also of the development of the "species" or "race": "The primary stage [of evolution] undoubtedly was bi-sexuality, such as still exists in the lowest classes of animal life and also during the first months of foetal existence in man" ([1886] 1965, 28); interference in the evolution of sexual types, in the form of hereditary degeneration or physical intervention, may result in "intermediary sexual gradations between the pure type of man and woman" (30). Krafft-Ebing shares the conviction, common at that period (Russett 1989, 131–50), that a strongly marked differentiation between the sexes is a product of advanced evolution and civilization, and that in "civilised races" the sexes are more different from each other than is the case among "less developed races": "The higher the anthropological development of the race, the stronger these contrasts between man and woman, and vice versa" (Krafft-Ebing [1886] 1965, 30). (Ellis, on the other hand, taking the unorthodox view that women are more highly evolved than men, claims that "modern civilisation is becoming . . . feminine" (Ellis 1904, 448) and that as men evolve they will become more like women.) Interestingly, Krafft-Ebing uses "mono-sexuality," a term which some bisexual theorists in the 1990s have reactivated to describe those whose orientation is towards one sex only, to describe the state of sexual-racial development which has evolved beyond the duality of bisexuality to a single-sexed harmony in which the sex predominating in the primary sexual characteristics corresponds with that predominating in the secondary: "The type of the present stage of evolution is mono-sexuality, that is to say, a congruous development of the secondary bodily and psychical sexual characteristics belonging to the respective sexual glands" (30).

Clearly, then, "bi-sexuality," which both Ellis and Krafft-Ebing see evidenced in such secondary characteristics as male nipples, facial hair in women and the clitoris as a bisexual analogue of the penis, is an important feature of the "heterosexual matrix" as it appears in *fin-de-siècle* sexology, arguably even its very heart: two sexes evolved from one originary bisexual state into a "mono-sexual" congruence between primary and secondary characteristics, analogous to the twentieth-century congruence between "sex" and "gender" identified by Butler. Moreover, the notion of "bi-sexuality" is what allows sexology to preserve a heterosexual logic even in cases of sexual inversion, characterized as "a mono-sexual psychic apparatus of generation, in a monosexual body which belongs to the opposite sex . . . a feminine psycho-sexual centre in a masculine brain, and *vice versa*" (Krafft-Ebing [1886] 1965, 427n.89) or as "a distinctly general, though not universal, tendency for [male] sexual inverts to approach the feminine type. . . . In inverted women a certain subtle masculinity or boyishness is equally prevalent. . . . Even in inversion the imperative need for a certain sexual opposition . . . still rules in full force" (Ellis 1897, 119–20)—this need for opposition sometimes leading, Ellis argues, to inter-racial sexual relationships among inverts, especially among women in American prisons. Moreover, the logic of bisexuality is a racial logic, both in the sense that it is a mark of evolution and its supersession by sexual dimorphism is, for Krafft-Ebing at least, a characteristic of racial superiority, and in the related sense that the two-sexed sexuality present in and developed from bisexuality is entirely organized around that principle of the propagation of "the race" which so thoroughly dominates the whole of human and animal life that neither procreative nor non-procreative sexuality is intelligible outside it.

European discourses of "race" at the *fin de siècle* were distinctively imperialist in character. According to Robert Young, "In the imperial phase, from the 1880s onwards, the cultural ideology of race became so dominant that racial superiority, and its attendant virtue of civilization, took over from economic gain or Christian missionary work as the presiding, justifying idea of the empire" (1995, 92). In fact all of the "justifying ideas" of empire mentioned by Young—economics, Christianity and racial superiority—had their corollaries in sexology; these are perhaps seen especially clearly in Krafft-Ebing. Krafft-Ebing's recurrent focus on masturbation, for example, as a primary factor in the development of perverse sexuality, is part of a tradition of seeing energy, particularly sexual energy and/or semen[6] itself, in economic terms, as something which may be "saved" or "spent" but which, if spent, should be spent wisely and not simply squandered. As Stephen Heath points out, the Victorian sexual imaginary mirrors the middle-class concerns of the day with commerce and economy, where *thrift* is the overriding principle for the regulation of both body and purse ("spending" is the term in common Victorian use as a colloquialism for orgasm): "*Thrift* is the supreme Victorian middle-class virtue: moderation, wise frugality, good housekeeping, the proper use of money and energy, ordered and regular expenditure; *unthrift* is thus waste, excess, ruinous expenditure, everything that is most immoral, a profound social disturbance."[7] Krafft-Ebing's repeated warnings against masturbation, which "despoils the unfolding bud of perfume and beauty" (189) and can lead young people into perversity if left unchecked, thus forms part of a more general sexual analogue of that "Protestant ethic" of thrift, frugality and asceticism which Weber was to identify as a key component of the "western" "spirit of capitalism" just as the final edition of *Psychopathia Sexualis* was appearing in print (Weber 1930).[8] More explicitly, Krafft-Ebing invokes the moral and spiritual superiority of Christianity over the beliefs of other "races" in his discussion of sexual development through history, with particular reference to Islam—significantly at a time when Germany had considerable economic and military interests in (among other places) Turkey. [. . .] One passage in particular, giving a detailed account of Krafft-Ebing's view of bisexuality and of its sexual and racial dynamics, is worth quoting at length:

> The author of this book has made an attempt to utilize facts of heredity for an explanation [of congenital homosexuality]. . . .
>
> All attempts at explanation made hitherto on the ground of natural philosophy or psychology, or those of a merely speculative character are insufficient.
>
> Later researches, however, proceeding on embryological (onto- and phylogenetic) and anthropological lines seem to promise good results. . . .
>
> . . . [T]hey are based (1) on the fact that bisexual organization is still found in the lower animal kingdom, and (2) on the supposition that monosexuality gradually developed from bisexuality. . . .
>
> *Chevalier* . . . proceeds from the original bisexual life in the animal kingdom, and the original bisexual predisposition of the human foetus. According to him the difference in the gender, with marked physical and psychical sexual character, is only the result of endless processes of evolution. The psycho-physical sexual difference runs parallel with the high level of the evolving process. The individual being must also itself pass through these grades of evolution; it is originally bisexual, but in the struggle between the male and female elements either one or the other is conquered, and a monosexual being is evolved which corresponds with the type of the present stage of evolution. But traces of the conquered

sexuality remain. Under certain circumstances these latent sexual characteristics . . . may provoke manifestations of inverted sexuality. . . .

If the structure of this opinion is continued, the following anthropological and historical facts may be evolved [*sic*]:

. . . This destruction of antipathic sexuality is at the present not yet completed. In the same manner in which the appendix in the intestinal tube points to former stages of organization, so may also be found in the sexual apparatus—in the male as well as in the female—residua, which point to the original onto- and phylogenetic bisexuality. . . .

Besides, a long line of clinical and anthropological facts favor this assumption. . . .

. . . Manifestations of inverted sexuality are evidently found only in persons with *organic taint*. In normal constitutions the law of mono-sexual development, homologous with the sexual glands, remains intact. . . .

The facts quoted seem to support an attempt of an historical and anthropological explanation of sexual inversion.

It is a disturbance of the law of the development of the cerebral centre, homologous to the sexual glands (homosexuality), and eventually also of the law of the mono-sexual formation of the individual (psychical "hermaphroditism"). In the former case it is the centre of bi-sexual predisposition, antagonistic to the gender represented by the sexual gland, which in a paradoxical manner conquers that originally intended to be superior; yet the law of mono-sexual development obtains.

In the other case victory lies with neither centre; yet an indication of the tendency of mono-sexual development remains in so far that one is predominant, as a rule the opposite. . . .

In the first case it must be assumed that the centre which by right should have conquered was too weak. . . .

In the second case both centres were too weak to obtain victory and superiority. (Krafft-Ebing [1886] 1965, 226–9)

A number of features appear in this passage which point towards the inherently racial nature of bisexuality for Krafft-Ebing. First, Krafft-Ebing is insistent that the explanation for sexual phenomena is to be sought in the fields of anthropology—the study of "race," with the characteristic ambiguity between "human race" and particular "races"—and embryology, thus reiterating the congruence of ontogeny with phylogeny and, at the same time, locating such phenomena firmly on racial ground. Second, an ambiguity appears in the meaning of "antipathic sexuality," a phrase widely used throughout the text as though it were simply interchangeable with "inversion" and "homosexuality" but which here is taken to refer to a physical, rather than psychosexual, phenomenon in the appendix-like persistence of male physiological features in all female bodies and vice versa, and which occurs in a passage (immediately following that quoted) in an apparently different sense again, as a subdivision of inversion rather than as a synonym for it: "The antipathic sexual instinct is only the strongest mark left by a whole series of exhibitions of the partial development of psychical and physical inverted sexual characters" (230). These slippages in meaning may perhaps be read as symptomatic of a discursive move which attempts both to universalize and to minoritize, to render certain phenomena both general and particular so that the power-effects of the discursive attention upon the latter are circulated and diffused throughout the former. The same is arguably true of those slippages, often noted but not explained by historians, in the meanings of the term "race": the slippages are not merely incidental to the power-effects of racial and sexual discourses, but are integral to those very effects, a crucial part of the production and regulation of sexual-racial subjects. Sexual abnormality, racial inferiority haunt the bodies even of those to whom the particularity of those conditions does not actually apply; no subject is exempt from sexual or racial interrogation.

Third, the dynamic of bisexuality itself is articulated here in the language of conquest and racial struggle. A conflict between opposing forces properly ends in the conquest of one side by another, but traces of the defeated force remain and may, under certain conditions, rise up to overthrow the rule of the conqueror. The imperialist logic, and the imperialist anxiety, are evident here, and homosexuality is clearly imagined as a rising up of an inferior force against its rightful ruler: precisely, a *racially* inferior force, since it is on the basis of the demands of "race" and procreation that Krafft-Ebing's entire theory of "bi-sexuality" and "mono-sexuality" is built, with bisexuality itself presented as "primitive." Moreover, in the case of men—a case which Krafft-Ebing in common with other writers then and since takes as paradigmatic—it is the *female* element which should be the conquered, the male which should conquer; and this resonates with the imperialist feminization of conquered peoples throughout the nineteenth and early twentieth centuries to which scholars of imperialism have recently drawn much attention, particularly in contexts of European colonial rule over peoples imagined as "Oriental," with connotations of exoticism, decadence and excess (de Groot 1989; Kabbani 1986; Said 1978)—recall Krafft-Ebing's comments on the inferiority of Islam to Christianity, specifically in relation to its sexual and marital practices. The point is that this imperial dynamic is not simply an accidental feature of Krafft-Ebing's conception of bisexuality. The development of mono-sexuality through conflict between opposing forces of male and female elements is the whole point of Krafft-Ebing's theory here. Taken together, it is the racial logic of dimorphic evolution and heterosexual procreation, and the imperial logic of conquest and overthrow, which constitute his conception of "bi-sexuality" as such. "Race" is the very stuff of bisexuality; in the sexual sphere, no sense can be made of the one without the other.

Unraveling the racial logics of sexuality is important and ongoing work, and clearly it raises questions about how today's sexual dissidents are to articulate their own identities as dissident while also paying attention to the racialized roots of those identities. The problematics of sexual-racial ethics is an area too large and complex to discuss adequately here. [. . .]

[. . .] An exclusive focus on the [hetero/homo] binary as if its dominance were simply a historical given—even as if it were the only important dynamic at work in sexuality at all—is a distortion of historical complexity; and foreclosure of any critical interrogation of bisexuality through such a focus also forecloses crucial questions about "race." [. . .] "Bi-sexuality" demonstrates the specific ways in which sexual categories are always also racial categories; in which the relationship between "sex" and "race" is not additive, but mutually constitutive; in which our contemporary sexual identities—which many sexual dissidents today understand and experience as progressive and liberatory—have historically been constituted as racialized identities. Dislodging the hetero/homo binary from its currently privileged position may force us to recognize that even the "outlaws" of sexual dissidence are unwitting legislators for the reproduction of "race."

Notes

1 This is perhaps most clearly to be seen in Freud's "Three Essays on the Theory of Sexuality" (1905), an early and seminal work located very much within the nineteenth-century sexological tradition even as it breaks away from it (Freud [1905] 1977).
2 Certainly not *all* sexual practices can be subsumed under the hetero/homo divide: when one masturbates, is that a heterosexual or a homosexual act? To be sure, "masturbator" has not thus far operated as a sexual identity in the twentieth century, even within those queer discourses which seek to reclaim otherwise "outlawed" sexualities, but the fact that it has not done so is a point that should be interrogated rather than simply foreclosed with the alibi of a totalitarian hetero/homo binary regime (Bennett and Rosario 1995).
3 Exceptions to this neglect have included (Dollimore 1991; Fuss 1995; McClintock 1995).
4 Second italic original, first and third italics mine [Editors' note: no italics in quotation].
5 The hyphenation of terms such as "bi-sexual" and "mono-sexual" is inconsistent in these texts, especially in the translation of Krafft-Ebing. However, the most common occurrence is of hyphenated "bi-sexuality," and I shall reflect that convention in the texts quoted.
6 The anxiety is strongly oriented around male bodies, with Krafft-Ebing in particular harping on the perils of masturbation for masculinity. Anxieties about female masturbation took a somewhat different form: see e.g. (Dijkstra 1986, esp. pp. 66–81).
7 See (Heath 1982, 18 cf. pp. 20–3) on spermatorrhea. (Birken 1988; Dijkstra 1986; Laqueur 1995; Russett 1989) on similar descriptions of neurasthenia in economic terms.
8 Weber's essays on the "Protestant ethic" first appeared in German in 1904–5.

Works Cited

Bennett, Paula, and Vernon A. Rosario, eds. 1995. *Solitary Pleasures: The Historical, Literary, and Artistic Discourses of Autoeroticism*. New York: Routledge.
Birken, Lawrence. 1988. *Consuming Desire: Sexual Science and the Emergence of a Culture of Abundance, 1871–1914*. Ithaca, NY: Cornell University Press.
Bland, Lucy. 1995. *Banishing the Beast: English Feminism and Sexual Morality, 1885–1914*. New York: Penguin.
Burton, Antoinette M. 1990. "The White Woman's Burden: British Feminists and the Indian Woman, 1865–1915." Women's Studies International Forum.
Butler, Judith. 1990. *Gender Trouble: Feminism and the Subversion of Identity*: Psychology Press.
Davin, Anna. 1978. "Imperialism and motherhood." History Workshop.

de Groot, Joanna. 1989. "'Sex' and 'Race'." In *Sexuality and Subordination Interdisciplinary Studies of Gender in the Nineteenth Century*, edited by Susan Mendus and Jane Rendall, vii, 260 p. New York: Routledge.
Dijkstra, Bram. 1986. *Idols of Perversity: Fantasies of Feminine Evil in Fin-de-Siècle Culture*. New York: Oxford University Press.
Dollimore, Jonathan. 1991. *Sexual Dissidence: Augustine to Wilde, Freud to Foucault*. Oxford: Oxford University Press.
Ellis, Havelock. 1897. *Studies in the Psychology of Sex*. Vol. Volume I: Sexual Inversion. London: Oxford University Press.
———. 1904. *Man and Woman: A Study of Human Secondary Sexual Characters*. London: Walter Scott Publishing Co.
Freud, Sigmund. [1905] 1977. "Three Essays on the Theory of Sexuality." In *Pelican Freud Library*. London: Pelican.
Fuss, Diana. 1995. *Identification Papers*. New York: Routledge.
Heath, Stephen. 1982. *The Sexual Fix*. 1st American ed. London: Macmillan.
Kabbani, Rana. 1986. *Europe's Myths of Orient*. Bloomington: Indiana University Press.
Krafft-Ebing, Richard von. [1886] 1965. *Psychopathia Sexualis, with Especial Reference to the Antipathetic Sexual Instinct*. Translated by Franklin S. Klaf. New York: Stein and Day.
Laqueur, Thomas Walter. 1995. "The Social Evil, the Solitary Vice, and Pouring Tea." In *Solitary Pleasures: The Historical, Literary, and Artistic Discourses of Autoeroticism*, edited by Jane Bennett and Vernon Rosario, 155–62. New York: Routledge.
McClintock, Anne. 1995. *Imperial Leather Race, Gender, and Sexuality in the Colonial Conquest*. New York: Routledge.
Rowbotham, Sheila, and Jeffrey Weeks. 1977. *Socialism and the New Life: The Personal and Sexual Politics of Edward Carpenter and Havelock Ellis*. London: Pluto Press.
Russett, Cynthia Eagle. 1989. *Sexual Science: The Victorian Construction of Womanhood*. Cambridge, MA: Harvard University Press.
Said, Edward W. 1978. *Orientalism*. 1st ed. New York: Pantheon Books.
Sedgwick, Eve Kosofsky. 1990. *Epistemology of the Closet*: University of California Press.
Smith-Rosenberg, Carroll. 1991. "Discourses of Sexuality and Subjectivity." In *Hidden from History: Reclaiming the Gay and Lesbian Past*, edited by George Chauncey, Martha Vicinus and Martin B. Duberman, 264–80. New York: Penguin.
Ware, Vron. 1992. *Beyond the Pale: White Women, Racism, and History*. New York: Verso.
Weber, Max. 1930. *The Protestant Ethic and the Spirit of Capitalism*. Translated by Talcott Parsons. London: G. Allen & Unwin.
Young, Robert. 1995. *Colonial Desire: Hybridity in Theory, Culture, and Race*. New York: Routledge.

From Masturbator to Homosexual

The Construction of the Sex Pervert

LADELLE McWHORTER

IN OUR POST-FREUDIAN ERA IT IS ALL TOO EASY TO ASSUME THAT sexuality was always a central issue in human societies and that racism was always and inevitably a product of sexual fantasy. [. . .] But what actually occurred, I believe, is something close to the reverse, namely, that sexuality as a unitary field of knowledge and a network of institutional power is to a great extent a creature of scientific racism—or, more conservatively, that sexuality's overwhelming significance and pervasive force in early twentieth-century Anglo-America is largely due to the ways in which the forces of scientific racism used it to fashion white supremacist strategies and management techniques.

[. . .] Scientific racists were prominent among those calling for an end to what came to be called Victorian sexual repression in the early part of the twentieth century. Scientific racists like Howard and Lydston insisted that sexuality be studied and publicly discussed; after all, the future of The Race was in the balance. Sexuality as the means of reproduction had to be safeguarded and nurtured. As the site where the forces of development were most likely to falter or go wrong, it also had to be diligently monitored. Most importantly, as the mechanism by which the forces of evolution would produce the future of humanity, sexuality had to be carefully and painstakingly shepherded and managed. Science would lead the

---

Excerpted from Ladelle McWhorter, "Scientific Racism and the Threat of Sexual Predation" in *Racism and Sexual Oppression in Anglo-America: A Genealogy*. Bloomington: Indiana University Press, 2009.

way. Biologists, physicians, sexologists, and psychiatrists would discover sexuality's laws and norms. Forensic and psychiatric experts would develop techniques to contain deviance. But then these scientists and technicians, along with educators and religious leaders, would have to train each and every Nordic man, woman, boy, and girl to recognize signs of trouble in themselves and others and to instill in them the self-discipline necessary to stay within the parameters of healthy functioning.

Although Foucault does not tell precisely the story that I will develop here—in fact he seems to give sexuality chronological precedence over scientific racism (e.g. Foucault 1978, 149)[1]—his history of sexuality can help us understand how sexuality came into existence as a scientific phenomenon in the nineteenth century and thus how it became available for incorporation into scientific racist projects. [. . .]

Foucault asserts that sexuality as a unitary field of knowledge and experience did not exist in the seventeenth century. That is not to say that there were no orgasms or pregnancies, no courtship rituals or condemnations of carnal sins. Of course there were—and a lot else besides—all of which eventually became part of what we now think of and experience as sexuality. Foucault's point is that what now seems to be a unitary phenomenon at the center of human life and personal identity was then fragmentary and dispersed into a variety of separate domains, many of which were peripheral to most individuals' daily life, health, and well-being. People didn't think of themselves as fundamentally sexual beings, as beings with a sexual orientation and a sexual identity that established them in their very selfhood. [. . .] When something went wrong in their relationships or emotional lives, they didn't presume that whatever was wrong necessarily had anything to do with their libido. By the end of the nineteenth century, however, people in the mainstream of modern culture in industrialized countries did think of themselves as fundamentally sexual beings. They did believe that the most mundane aspects of their feelings and behavior as well as their bodily health were deeply connected to their sexual natures, so they thought that when something went wrong for them, it probably did have something to do with their sexual desires or practices. They also thought that when something went wrong with other people—when they behaved in seemingly irrational or unusual ways, when they fell ill, when they expressed feelings different from the norm—something might be wrong with their sexuality. Sexuality had become an object of knowledge by the end of the nineteenth century, and a very precious object of knowledge at that. Knowing one's own sexuality and understanding the sexuality of others—especially abnormal others—had come to seem crucial to getting along in the world. [. . .]

[. . .] [I] want to suggest some of the ways in which [Foucault's treatment of the "movements" that produced sexuality as we know it] articulated with the discursive and institutional practices of scientific racism to give rise to the figure of the feebleminded or weak-willed sexual predator in all his/her guises: the black rapist (and erstwhile werewolf), the alluring syphilitic whore, the sex-crazed imbecile, the conniving female sexual invert poised to recruit, and the homosexual child molester. Scientific racism, with its intense fear of genetic corruption, created this figure; and in it—long after scientific racism's official demise and, indeed, throughout the twentieth century down to the present day—its racist preoccupations and presumptions live on.

The direct ancestor of the late nineteenth century's sexual predator was the early nineteenth century's masturbator. The masturbator—a man, woman, or child consumed by deviant desire and addicted to the practice to the point of death—made a rather sudden appearance in European history in the eighteenth century. Prior to that time masturbation was officially condemned as a sin because it made use of the generative organs to produce a sterile pleasure, but it was not thought to pose a medical threat. Then suddenly everything changed. [. . .]

[. . .] According to the theory of organic degeneration that gained currency in the 1860s, the morbid somatic effects of masturbation were not confined to the present generation, and even limited indulgence could have a compounding effect. Robert Nye (1985) sums up the problem:

> In the absence of some countervailing external force, the syndrome developed an autonomous hereditary momentum, exhibiting its advance in worsening behavior and physical signs. The weakened capacity for "resistance" made the individual organism vulnerable to disease and hostile environments. The "moral" effects expressed themselves as will pathologies, that is as a catastrophically reduced ability to resist "impulsions" of instinct, the blandishments of sensual allure, the wine shop, or easy money. (59–60)

White men and women who masturbated, it would seem, ran the risk of becoming just like newly emancipated Negroes, unable to see the consequences of their actions or resist the allure of sources of immediate gratification. Masturbators who managed to reproduce despite their debility would pass their weakness on to their progeny in ever more concentrated form, bringing forth children who were alcoholics or epileptics and grandchildren who were idiots or homicidal maniacs. Just as the bloodlines of Indians and Negroes were declining toward extinction, white bloodlines could be corrupted by masturbators and brought to a similar evolutionary end. Survival of the fittest meant survival of the strong-willed, and masturbation was both a symptom and a cause of weak will. [. . .]

Thus elevated to the status of a threat to the continuation of the Race, a sensual pleasure that had little or nothing to do with procreation or physical health or heredity became a focal point for all kinds of therapeutic and pedagogical intervention. And those interventions were tolerated—in fact, demanded—by the educated public, because masturbation had been connected scientifically with concerns about health, procreation, and heredity through the emerging concept of "sexuality." Masturbation produced not just nervous excitation or debility but, precisely, a disturbed or corrupted sexuality. Thus it affected all aspects of that sexuality, including what we might call a person's gender (gait, gesture, vocal tone, dress, hobbies, career interests, and so on) and appearance and civility, as well as his or her sense of self, familial relationships, friendships, and ability to procreate.

Preventing masturbation among the better classes of white people was crucial if the human race was to survive. Thus the project of stopping it grew into the enormous task of managing upper and middle-class childhood. [. . .] Masturbation could undermine Nordic masculinity by producing a nervous exhaustion that rendered a man indifferent to the attractions of the opposite sex. Realizing he could not achieve erection in the presence of a female, the masturbator developed a dread of what would eventually be called "heterosexual encounters." Beard called it a "dread of intercourse" ([1898] 1972, 106). Meanwhile, masturbation inevitably heightened desires even as it diminished normal physical capacities. The masturbator required more stimulation to achieve satisfaction, and soon simple self-abuse would fail to gratify. He began to crave ever more perverse activities, and, in his "dread of intercourse," he turned to his own sex. Thus the masturbator eventually would become an insatiable effeminate pervert.[2]

In the worst cases, these perverts became delusional. They took on women's names and dress. Some even insisted that they really were women. This was monomania, a form of insanity that was incurable. In less advanced cases, the pervert understood that he was perverted, although he might enjoy his perverse indulgences so much that he did not want to be cured. In fact, Beard believed most sex perverts fell into this category: "Cases of sexual perversion are very much more frequent than is supposed; but they are rarely studied by scientific men, and only in exceptional cases do they consult scientific men. This class of people do not wish to get well" (101). Beard compares them to opium eaters on this point.

Masturbating women were subject to this same pattern of indifference, fear, and perversion. As James Kiernan writes, "The female masturbator of this type usually becomes excessively prudish, despises and hates the opposite sex, and frequently forms a furious attachment for another woman, to whom she unselfishly devotes herself" (1888, 172). Kiernan links this behavior to necrophilia and vampirism. But the outcome of masturbation was not simply uncontrollable perverse desire and self-debasement in ignoble practices. The effects in both sexes were constitutional, according to Beard: "The subjects of these excesses go through the stages of indifference and of fear, and complete the circle; the sex is perverted; they hate the opposite sex, and love their own; men become women, and women men, in their tastes, conduct, character, feelings, and behavior" (1972b, 106). Through masturbation, individuals virtually changed their sex. What greater threat could there be to Nordic masculinity?

Obviously, for men this was an evolutionary step backward, and thus it was a loss not only for the individual but also for the entire race in its journey toward world domination. We might imagine that for women, though, sexual "inversion" was an evolutionary step forward, since men were held to be higher on the evolutionary scale than women of any race. But not so. In fact, of course, these masturbating monsters did not actually rid themselves of their original anatomical sex before taking on the traits of the other; consequently, they merely blurred the lines between the sexes in their conduct and physiology and thus became a sort of hybrid or third sex. This blurring was itself degenerate, for sexual differentiation was widely believed to a product of advanced evolution. Krafft-Ebing asserts this as a matter of stage setting for his *Psychopathia Sexualis:* "The secondary sex characteristics differentiate the two sexes; they present the specific male and female types. The higher the anthropological development of the race, the stronger these contrasts between man and woman, and vice versa" (Krafft-Ebing [1886] 1965, 28).[3]

Similar comments abound in biological, sexological, and medical literature. Savages were much less sexually differentiated than civilized Victorian ladies and gentlemen, the experts asserted. Native American females, for example, had coarse features and physical strength that approached the masculine. The same could be said of African females and American Negresses. Even in somewhat more advanced races—such as the Chinese and the Jew—sexual difference was less apparent than in the refined Nordic race (Magubane 2003, 108–10; McClintock 1995, 52–55; Mosse 1985, 143–46). Kiernan employed this common belief in his study of sexual perversion [. . .]:

> The original bi-sexuality of the ancestors of the race, shown in the rudimentary female organs of the male, could not fail to occasion functional, if not organic, reversions when mental or physical manifestations were interfered with by disease or congenital defect. The inhibitions on excessive action to accomplish a given purpose, which the race has acquired through centuries of evolution, being removed, the animal in man springs to the surface. Removal of these inhibitions produces, among other results, sexual perversions. (1888, 129)

Once the patient's will-power or reason was compromised by masturbation or degenerative disease, "reversion" to the primordial bestial type would be the result. Female inverts, therefore—with their husky voices, educational ambitions, and incredibly enlarged clitorises—were just as evolutionarily retrograde as their mincing male counterparts, and in their lusty, predatory pursuit of weak-willed white woman-flesh, they were every bit as dangerous to the future of the Nordic race.

The slide from masturbation to homosexuality seems bizarre from a twenty-first-century perspective. However, that is partly because current definitions of masturbation are very narrow compared to the definitions operative in the nineteenth century. We think of masturbation as self-stimulation only, accomplished with the hand or perhaps with an object held in the hand. But consider this textbook definition of masturbation from 1896: "venereal orgasm by means of the hand, the tongue, or any kind of body by one's self or another person" (Gibson 1997, 116). The war on masturbation was not, in fact, confined to the "solitary sin"; it included attacks on what we would call mutual masturbation and oral sex. By mid-twentieth-century standards, some nineteenth-century "masturbators" subject to these dramatic social and medical interventions were not actually masturbating; they were committing homosexual acts and would have been considered homosexuals. By nineteenth-century definitions, however, they remained onanists, not inverts, until they graduated from mutual masturbation, fellatio, or cunnilingus to anal intercourse or tribadism, or when their appearances or manners were judged to be gender-transgressive.

There were two categories of inverts. First, there were those whose condition was a result of self-induced degeneracy through willful vice. These despicable individuals should be punished to the full extent of the law unless they had already passed into the stage of incurable monomania. However, increasingly influenced by the personal disclosures of inverts themselves, many nineteenth-century physicians began to believe there was a second group. George Beard [. . .] [and] Krafft-Ebing noted that some inversion appears to be congenital and that the degeneracy that produces sexual inversion is heritable (Beard [1898] 1972, 107; Krafft-Ebing [1886] 1965, 188). Studies of hermaphrodites demonstrated that male and female sex organs could sometimes be mixed together in a single body. We know that some people are born with the gonads of one sex but genitalia characteristic of the other, many physicians reasoned, so maybe some people are born with the gonads and genitalia of one sex but the brain and neurological system of the other. They may look like normal males or normal females, but neurologically they are hermaphrodites.[4]

Whether congenital or acquired, if degeneration was so far advanced that monomania had set in, lifelong confinement was the only course of action that made any sense. It wouldn't do to have monomaniacal sexual inverts running around loose, especially with a population of fragile white people in the throes of a difficult evolutionary advance that many were ill-prepared to negotiate. But it might not be fair to punish congenital inverts, many physicians and sexologists believed, because their actions were not truly voluntary. As James Kiernan put it, "There can be no legal responsibility where free determination of the will is impaired" (1892,

185). Congenital inverts were naturally weak of will, lacking in "nervous force," unable to resist the perverse urges that their degenerate condition aroused. Such individuals might undergo episodic periods of organically produced sexual furor during which they were entirely devoid of self-control. [. . .] The question was how to identify these individuals before they did any damage and eliminate the danger they posed without compromising the justice system by punishing people whose actions were totally involuntary. Thus began an entire neuropsychiatric industry—data collection and classification leading to establishment of signs and procedures by which forensic experts could recognize a genuine sex pervert when they saw one.

This was especially important because when the law took hold of sexual inverts without sound psychiatric advice, judges were apt to impose a fixed prison sentence and be done with the matter. At the end of the sentence, however, a congenital invert would still be a congenital invert. If released, he or she would simply "prey on society again." Psychiatrists had a better, more scientifically informed solution: Persons "mentally and sexually degenerate from the first, and therefore irresponsible, must be removed from society for life" (Kiernan 1892, 335). They should not be stigmatized as criminal or subjected to punishment; they simply needed lifelong psychiatric care, and society needed protection from their morbid influence.

Neurologists and psychiatrists both in Europe and in the United States tended to agree on this point. Edward Mann, medical superintendent of New York's Sunnyside Sanitarium for Diseases of the Nervous System, held that many cases of sexual inversion were congenital atavisms: "There is very often a true congenital moral deprivation with strong animal propensities, which makes a person practically insane from birth. . . . There is an entire perversion of the moral principle and there are no good or honest sentiments" (1893, 272).[5] Mann's definition of insanity was more commodious than most, but his practical conclusion was the same: a penitentiary stint cannot reform these people. [. . .] Furthermore, these insane sex perverts should be institutionalized because they can induce perversion in sane but weak-willed or immature individuals and will do so if given half a chance (274). G. Frank Lydston (1904) concurs: "All incurable victims should be permanently removed from our social system. They are sources of moral contagion and promoters of sexual crime to whom the right to remain in society should be denied" (421).

That is to say, sexual inverts recruit. As Krafft-Ebing noted, many inverts whose condition is acquired rather than congenital got the way they are, not merely by making depraved, self-indulgent choices (which they did, and for which we must condemn them), but also by becoming the prey of congenital inverts. As so many physicians of the time pointed out, middle- and upper-class white men, with their newly evolved and relatively weak hereditary traits for civility and their highly refined, tightly integrated nervous systems, were very fragile creatures. In their high-stress positions as the captains of industry and the inventors of the future of mankind, they were really quite vulnerable. Especially in youth and young bachelorhood, they might succumb to the influences of hardened, manipulative inverts and engage in mutual or oral masturbation. Without intervention, the situation could escalate. They could lose their manhood and be lost to the Race. Perhaps even more frightening was the specter of the female invert preying upon the delicate flower of white womanhood, offering not only the attention and caresses her suitors or young husband might not have the time to provide but also dubious opportunities for excitement such as intellectual conversation, a college education, or a serious role in a movement for political or social reform. Sexual inverts were sexual predators; for the sake of the future of the Race, they had to be stopped.

Although there were dissenters, many sexologists believed that white female inverts were particularly intelligent and cunning. They had masculine brains, after all—not quite white male brains, but brains like those of males of lower races, the Chinese for instance (Gibson 1998). Like male inverts, mental defectives, and savages, they also had heightened sex drives. "The sexual life of individuals thus organized manifests itself, as a rule, abnormally early, and thereafter with abnormal power" (Krafft-Ebing [1886] 1965, 223). Female inverts of all races were usually classified as either nymphomaniacs or erotomaniacs (Aldaraca 1995). Either way, like all black men and women, they were sexually insatiable. When they found a white woman who was neurasthenic or suggestible, they would not hesitate to entice her into a sexual relationship that would drag her down the path on which their own primitive or degenerate natures had already set them.

A surprising number of white women were vulnerable to same-sex seduction, according to the experts. That they didn't often succumb had more to do with external circumstances than with their own potentials and inclinations. Krafft-Ebing believed that most white women who began life with a tendency toward inversion (at least those of the middle and upper classes) were saved from expression of it by the constraints of Victorian feminine education. [. . .] This fact explained the high frequency of frigidity in married women, he believed; frigid women were latent inverts, saved from degenerate (but satisfying) sexual expression by the strictures of Victorian upbringing. All would be well unless such a lady came into contact with an irrepressible invert who appealed to her dark side. In that event, Krafft-Ebing warned, "we find situations analogous to those which have been described as existing in men afflicted with 'acquired' antipathic sexual instinct."[6] [. . .]

[. . .] In sum, then, girls who masturbate, girls who are sequestered from male attention, girls who are afraid to have sex for fear of pregnancy, wives whose husbands are poor lovers, and women whose male partners force them to do disgusting things are likely to turn willingly to any female invert in the vicinity. No wonder Victorian physicians feared the lurking presence of the atavistic, Chinaman-brained white female invert (Gibson 1998, 5)! Virtually every girl and woman they knew was a latent case of inversion just waiting to happen!

With these strokes, by the last decade of the nineteenth century, psychiatrists and neurologists painted the portrait of the homosexual predator, both male and female. This person was degenerate, sexually insatiable, in some descriptions insane and in some savagely atavistic, but in all cases not governed by reason or moral principle, and able to pass his or her condition to others through both heredity and enticement analogous to infection. Obviously such a person was a threat to the biological integrity of the Race and to the continued evolution of Civilization. Equally obviously, the threat posed was basically the same as that posed by the menacing imbecile and the savage Negro in the throes of his or her periodic *furor sexualis*. What was at stake was the purity of the Race and its fitness for survival. All these sexual predators were vectors of genetic pollution, conduits of abnormality and defect, pipelines for impurity. They all had to be neutralized. [. . .]

It is hardly necessary to assert that the most pervasive image of the homosexual in our culture is that of the sexual predator—the lurking, child-molesting, virgin-corrupting, disease-spreading pervert. We all live with that image even if we don't subscribe to it. We may have laughed when Matthew Shepard's murderers attempted to apply that label to their five-foot-two-inch, 105-pound victim, but we recognized the ploy, and we knew that many Americans would believe it. For many Americans believe that all homosexuals are sexual predators even if they are too small, weak, out-gunned, and out-numbered to protect themselves against their intended victims' outrage. Perhaps the only way to rid ourselves of this image's influence is to see it in the context of its history alongside the images of the imbecilic sex criminal, the black rapist, and the Jezebel or syphilitic, feebleminded whore.

Although these latter figures arose in disparate social, political, and professional contexts, they are remarkably similar. These creatures are, either continually or episodically, outside the governance of reason. Their affliction in every case is a matter of development—either faulty, arrested, or retrograde. They cannot be assimilated to society both because they cannot manage their own behavior well enough to function within its civil constraints and because they pose a biological threat to it in the form of contagion and corruption of germ plasm. The predatory homosexual, whether male or female, is their cousin, formed in the same lineage and carrying the same taints. For the sake of national security and the future of the human race and civilization as we know it—in other words, in the name of Anglo-Saxon world domination—all these people had to be segregated from the rest of the population. If they were outside our national boundaries, we had to close our borders to them. If they were already inside and we needed their labor, we either crowded them into ghettos or prisons or confined them in work camps and warehouses misnamed "asylums." [. . .]

It was a racial dream inspired by three hundred years of technological innovation and imperialist conquest and suffused with a science that took development to be the foundation and meaning of life. As a racial dream, a dream that was to be made reality through the willpower and work of the dreamers, sexuality was its primary tool. Along with genocide, sexuality was the main medium through which populations, races, the Race could be shaped.

That science lost its status as truth in the 1930s [. . .] but the dream lived on. So did the mechanisms and alliances that had been put in place during that great rush of Anglo-Saxon self-assurance: the carceral system and all its peripheral apparatus of surveillance, the sex-saturated nuclear family, a social welfare system that was all about the welfare of "society" and not at all about the welfare of the weak, the poor, or the disabled. By that time, three generations of Americans—of all descriptions—had been taught the tenets of white supremacy from the cradle. Their teachers had been scientists, physicians, scholars and educators, civic leaders, clergymen, even presidents. In the process, they had been taught that sexuality—procreation and heredity, public health, and child rearing and family life, as well as gender roles, bodily pleasures, and personal identity—lay at the base of all that they held or should hold dear and that the world as they knew it might come to an end if they or those around them were to deviate from prescribed sexual norms.

The science receded. But the husks and shells it left in place—the armor and weaponry it had constructed for itself to aid in its advance—remained, ready to be donned and wielded by anyone who could maneuver into position. Racism, even stripped of its scientific logic, still presented a formidable front, as John Lewis and Hosea Williams and Amelia Boynton Robinson well knew, standing on the Edmund Pettus Bridge in 1965. Scientifically grounded or not, it could still exercise a profound influence over all aspects of public and personal life. And it could still be deadly. The next two chapters [of *Racism and Sexual Oppression in Anglo-America*] show the ways in which its effects extended through the twentieth century and into the twenty-first in the lives of all Americans, but especially in the lives of those deemed abnormal.

Notes

1 This prioritization of sexuality over race is one that Brady Heiner (2007, 336) has also noted and for which he has criticized Foucault [. . .].

2 In this claim, Beard echoed his German predecessors, Johann Ludwig Casper [. . .] and [Henrich] Kaan [. . .] (Hekma 1994, 215–17).

3 *Editors' note:* Author annotated the reference to *Psychopathia Sexualis* as follows: "This edition was originally published in German in 1903 and was the last one Krafft-Ebing edited himself." All other authors in this volume who cite Krafft-Ebing's *Psychopathia Sexualis* use this same edition.

4 This argument has been revived by biologist Simon LeVay (1993).

5 *Editors' note:* The original citation actually reads (1892, 272). Editors assumed based on context that this reference was to Mann.

6 Merl Storr (1998, 20) points out that in this passage Krafft-Ebing basically undercuts his own distinction between congenital and acquired inversion. His analysis provides no way to tell the difference in most women.

Works Cited

Aldaraca, Bridget A. 1995. "On the Use of Medical Diagnosis as Name-Calling: Anita F. Hill and the Rediscovery of "Erotomania."." In *Black Women in America*, edited by Kim Marie Vaz, 206–21. London: Sage Publications.

Beard, George Miller. [1898] 1972. *Sexual Neurasthenia (Nervous Exhaustion): Its Hygiene, Causes, Symptoms and Treatment: With a Chapter on Diet for the Nervous*. 5th ed. New York: Arno Press.

Foucault, Michel. 1978. "The History of Sexuality, Vol. 1: An Introduction." *Trans. Robert Hurley. New York: Pantheon.*

Gibson, Margaret. 1997. "Clitoral Corruption: Body Metaphors and American Doctors' Construction of Female Homosexuality, 1870–1900." In *Science and Homosexualities*, edited by Vernon A. Rosario, 108–32. New York: Routledge.

———. 1998. "The Masculine Degenerate: American Doctors' Portrayals of the Lesbian Intellect, 1880–1949." *Journal of Women's History* 9 (4): 78–103.

Heiner, Brady Thomas. 2007. "Foucault and the Black Panthers 1." *City* 11 (3): 313–356.

Hekma, Gert. 1994. "'A Female Soul in a Male Body': Sexual Inversion as Gender Inversion in Nineteenth-Century Sexology." In *Third Sex, Third Gender: Beyond Sexual Dimorphism in Culture and History*, edited by Gilbert H. Herdt, 213–39. New York: Zone Books.

Kiernan, James G. 1888. "Sexual Perversion and the Whitechapel Murders." *Medical Standard* 4 (129–30): 170–72.

———. 1892. "Responsibility in Sexual Perversion." *Chicago Medical Recorder* 3 (May): 185–210.

Krafft-Ebing, Richard von. [1886] 1965. *Psychopathia Sexualis, with Especial Reference to the Antipathetic Sexual Instinct*. Translated by Franklin S. Klaf. New York: Stein and Day.

LeVay, Simon. 1993. *The Sexual Brain*. Cambridge, MA: MIT Press.

Lydston, G. Frank. 1904. *The Diseases of Society: The Vice and Crime problem*. Philadelphia: J.B. Lippincott.

Magubane, Zine. 2003. "Simians, Savages, Skulls, and Sex: Science and Colonial Militarism in 19th Century South Africa." In *Race, Nature, and the Politics of Difference*, edited by Donald S. Moore, Jake Kosek and Anand Pandian, viii, 475 p. Durham: Duke University Press.

Mann, Edward C. 1893. "Medico-Legal and Psychological Aspect of the Trial of Josephine Mallison Smith: Tried for Murder, in Philadelphia, Penn., November 29th, 30th, 31st and December 1st, 1892." *Alienist and Neurologist* (14):467–77.

McClintock, Anne. 1995. *Imperial Leather Race, Gender, and Sexuality in the Colonial Conquest*. New York: Routledge.

Mosse, George L. 1985. *Nationalism and Sexuality: Respectability and Abnormal Sexuality in Modern Europe*. 1st ed. New York: H. Fertig.

Nye, Robert A. 1985. "Sociology and Degeneration: The Irony of Progress." In *Degeneration: the Dark Side of Progress*, edited by J. Edward Chamberlin and Sander L. Gilman, 49–71. New York: Columbia University Press.

Storr, Merl. 1998. "Transformations: Subjects, Categories and Cures in Krafft-Ebing's Sexology." In *Sexology in Culture: Labelling Bodies and Desires*, edited by Lucy Bland and Laura L. Doan, 11–25. Chicago: University of Chicago Press.

"An Unnamed Blank that Craved a Name"

A Genealogy of Intersex as Gender

DAVID A. RUBIN

> The fact that my gender has been problematized is the source of my intersexual identity.
>
> CHASE, "AFFRONTING REASON"

IN "THE MEDICAL CONSTRUCTION OF GENDER: CASE MANAGEMENT of Intersexed Infants," Suzanne J. Kessler (1990) focalized a practice that was, up until the early 1990s, rarely discussed outside of specialized medical circles: the surgical normalization of infants born with sexual anatomies deemed to be nonstandard. Analyzing interviews with physicians and the medical literature on intersex treatment, Kessler argued that "members of medical teams have standard practices for managing intersexuality that ultimately rely on cultural understandings of gender" (4). In making this claim, Kessler emphasized the significance of clinicians' reliance on what is known as the optimal gender paradigm. Developed by psychoendocrinologist John Money and his various colleagues over the years, the optimal gender paradigm is a treatment model that seeks to help physicians select the most optimal gender for individuals born with atypically sexed anatomies. Its central presumption is that surgical normalization

---

Originally published in a slightly different form as David A. Rubin, "'An Unnamed Blank That Craved a Name': A Genealogy of Intersex as Gender," *Signs: Journal of Women in Culture and Society* 37, vol. 37, no. 4: 909–933.

can and should be used to foster the development of conventional gender identities. Kessler, however, was concerned about the ethical implications of this paradigm, specifically the ways in which it medicalized intersexual difference so as to maintain the gender order status quo. Noting that the vast majority of intersex "conditions" pose little or no health risk, she concluded that intersexuality "is 'corrected' not because it is threatening to the infant's life but because it is threatening to the infant's culture" (25).

Following Kessler's lead, during the past twenty five years a small but growing number of scholars have made vital contributions to feminist and queer theory, science studies, bioethics, medical sociology, and debates about human rights and bodily integrity by showing that intersexuality challenges naturalized understandings of embodiment through analyses of the medical construction of sexual dimorphism (Fausto-Sterling 1993; Fausto-Sterling 2000; Hird 2000; Preves 2003; Karkazis 2008; Davis 2015). As my language indicates, I am interested in the implications of an unremarked discursive shift that began to manifest itself as this body of interdisciplinary research developed. In Kessler's wake, the analytic preoccupation of intersex studies was displaced almost immediately from gender to sex, as evidenced by the titles of works published following "The Medical Construction of Gender" such as *Sexing the Body* (Fausto-Sterling 2000), "Sexing the Intersexed" (Preves 2002), and *Fixing Sex* (Karkazis 2008). One could interpret this shift as a transition from the social back to the somatic, reading the emphasis on *sexing* as consonant with and influenced by the recognition of the limitations of the essentialism/constructionism divide and the consequent push to rethink the materiality of the body in 1990s feminist theory post-*Gender Trouble* (Butler 1990, 1999). But this alone does not explain why gender receded into the theoretical background of intersex studies as the field began to congeal. While scholars have undoubtedly sharpened critical perspectives on the medical and social treatment of people with intersex embodiments, their accounts have largely focused on rethinking the sex side of the sex/gender distinction. For this reason, less attention has been paid to questions about the genealogical relation between intersex and gender, questions that were implicitly posed but not fully answered in Kessler's initial *Signs* essay, such as: What is the historical relationship between intersex and the sex/gender distinction? How has the sex/gender distinction shaped and been shaped by intersex?

The elision of these questions has been reinforced by an influential strain of intersex activism. Throughout the 1990s and 2000s the Intersex Society of North America (ISNA) avowed that "intersexuality is primarily a problem of stigma and trauma, not gender." As Iain Morland argues, this claim crucially "acknowledged that affected individuals—rather than their parents or doctors—are experts on their own genders," and further suggested "that traditional treatment . . . often inadvertently creates trauma and thus fails by its own standards" (2011, 156–57). In this way, ISNA challenged the medical model of intersex management, and that model's surgical equation of dimorphic genitalia with normative sex, promoting instead a patient-centered approach founded on intersex adults' critical reflections on their experiences of medicalization. Although this claim buttressed ISNA's opposition to nonconsensual genital surgery, it also obscured and, due to ISNA's lasting impact, continues to obscure the powerful role of gender in the development of modern intersex medicine and the sciences of sexual health more broadly. Before it became a key term in feminist discourse, before it came to signify the social construction of femininity and masculinity, and before it became an assumed core of modern personhood as such, gender was formulated in mid-twentieth century American sexology as a diagnostic solution to the so-called medical emergency of intersex bodies, or bodies in doubt (Reis 2009). The story of intersex is not only, as ISNA asserted, a story about "shame, secrecy, and unwanted genital surgeries," a story about "stigma and trauma" but also a story about the regulation of embodied difference through biopolitical discourses, practices, and technologies of normalization that materialize in, through, and as gender. [. . .]

As Jennifer Germon argues, gender does in fact have a history, and "a controversial one at that" (2009, 1). Germon cites Bernice Hausman (1995), arguing that it was not until the mid-twentieth century that English speakers began using gender as an ontological category, a category said to denote masculine and feminine states of subjective being. In particular, Germon suggests that Money's influence upon the career of the gender concept has been even more decisive than Kessler initially indicated. According to Germon, it was through Money's research that the gender concept came to be recognized as an explanatory measure of human behavior in the biomedical and social sciences. In addition, Germon argues that Money's ideas, despite their problematic investments in medical paternalism and the binary model of sexual difference, nevertheless manifest a strong interest in understanding nature and culture within a more complex interactionist framework.

In their recent book *Fuckology: Critical Essays on John Money's Diagnostic Concepts*, Lisa Downing, Iain Morland, and Nikki Sullivan (2015) challenge Germon's optimism that Money's work adopts an interactionist approach to the relationships between the cellular, environmental, and experiential domains. Downing, Morland, and Sullivan show that Money's claims were often conflicting, self-undermining, and dysfunctional. For instance, Sullivan contends that Money was neither an essentialist nor a constructionist in any simple or straightforward sense, but she does suggest that Money's model of gender "posits the biological as foundational" to the subsequent development of psychosexual gender identities and roles (Sullivan 2015, 20). Doing critical justice to Money's vast and contradictory body of research and its legacies, she concludes, requires that we "trouble the tendency to see in dimorphic terms" (20).

My analysis converges with Downing, Morland, and Sullivan's and Germon's in exploring the enduring significance of Money's research, and the centrality of the intersexed to the history and politics of gender. In an effort to deepen and extend these analyses, I argue not only that intersexuality played a crucial role in the invention of gender as a category in mid-twentieth century biomedical and, subsequently, feminist discourses; and that Money used the concept of gender to cover over and displace the biological instability of the body he discovered through his research on intersex; but also that Money's conception of gender produced new technologies of psychosomatic normalization. In contrast with Germon (2009), my aim is not "to critically reinvigorate Money's gender" concept (3) but rather, following Downing, Morland, and Sullivan, to more fully excavate the broad swathe of its regulatory power.

Gender in Money's Research

As the inventor of the term *gender role*, Money's work brings into focus the role of intersex as an origin of *gender* and of the sex/gender distinction. Indeed, as I will suggest, thirty-five years before *Gender Trouble*, Money posited gender as *prior* to sex. [. . .]

While studying the relation between endocrine functions and psychological states of hermaphroditism at Harvard in the 1950s, Money coined the term *gender role* as a diagnostic category and treatment protocol for patients whose anatomical configurations were regarded as unintelligible within the dominant frame of dimorphic sex. For people with intersex characteristics, whose bodies Money read as improperly sexed, *gender role* became a way for Money to predict and, as we will see, to literally fashion the sex they were "supposed" to have all along. Money's typical scientific approach used the abnormal to find and define the normal. His work on intersex helped to popularize the view that gender is central to the sexual health of persons in general.

Money first made reference to his theory of gender in a 1955 article published in the *Bulletin of the Johns Hopkins Hospital* titled "Hermaphroditism, Gender and Precocity in Hyperadrenocorticism: Psychologic Findings." In that paper, Money would later write in a 1995 retrospective essay on his life's work, "the word *gender* made its first appearance in English as a human attribute, but it was not simply a synonym for *sex*. With specific reference to the genital birth defect of hermaphroditism, it signified the overall degree of masculinity and/or femininity that is privately experienced and publicly manifested in infancy, childhood, and adulthood, and that usually though not invariably correlates with the anatomy of the organs of procreation" (18–19).

In their influential textbook *Man & Woman, Boy & Girl,* Money and Anke E. Ehrhardt offer a more general theory of hermaphroditism, claiming that the terms hermaphroditism and intersex can be used interchangeably as both "mean . . . that a baby is born with the sexual anatomy improperly differentiated. The baby is, in other words, sexually unfinished" (1972, 5). Two presuppositions ground this claim: first, that sexual anatomy has a proper mode of differentiation that, second, constitutes a complete or finished form of sexual dimorphism. [. . .] Money and Ehrhardt's understanding of intersex was not only pathologizing but was also structured by a spatial and temporal logic of human development whose telos is wholeness. As several critics have pointed out, this perspective is problematic in terms of its heteronormative and sexually dimorphic ideological biases (Chase 2002, Fausto-Sterling 2000, Holmes 2008).

These presuppositions were evident in Money's work from the start. Money first became acquainted with hermaphroditism in the Harvard psychological clinic, where he wrote his PhD dissertation on "Hermaphroditism: An Inquiry into the Nature of a Human Paradox" (1952). For his dissertation, Money conducted 10 case studies with interviews and collected 248 cases from a medical literature review to show that "psychosexual orientation bears a very strong relationship to teaching and the lessons of experience and should be conceived as a psychological phenomenon" (7). By "psychosexual orientation," Money meant "libidinal inclination, sexual outlook, and sexual behavior" (5). In "Lexical History and Constructionist Ideology of Gender" Money quotes his dissertation at length to reveal how his studies of hermaphroditism generated for him the following problem: "For the name of a single conceptual entity, there are too many words in the expression 'libidinal orientation, sexual outlook, and sexual behavior as masculine or feminine in both its general and its specifically erotic aspects.' The challenge to give a unitary name to the concept embodied in these many words became pressing after my case load of hermaphrodites studied in person had, after 1951, expanded from ten to sixty in Lawson Wilkins' Pediatric Endocrine Clinic at the Johns Hopkins Hospital, at which time a concise report of the findings became essential" (1995, 20). Studying individuals with anatomical configurations he regarded as anomalous, Money initially and inadvertently proliferated diagnostic categories; his research generated, he says, "too many words." This excess of signification highlights the degree to which intersexuality troubled the symbolic resources of Money's biomedical episteme. To overcome the discursive proliferation that his studies of intersexuality inaugurated, Money went in search of "a unitary name." In short, Money sought to establish an exhaustive, monolithic taxonomy to explain and contain the discursive excess generated by hermaphroditism.

Money's dissertation suggested that psychosexual orientation is shaped by social and psychological factors, and in forwarding this thesis Money was staging an argument with previous psychologists and sex researchers who held that psychosexual orientation was biological and innate. In the 1950s, a time when biological determinism, while contested, was still dominant in the hard sciences (Meyerowitz 2004), Money's insistence that masculinity and femininity could not be reduced to biology alone remains quite remarkable. Summarizing his post-1951 findings, Money explains in "Lexical History and Constructionist Ideology of Gender" that:

> The first step was to abandon the unitary definition of sex as male or female, and to formulate a list of five prenatally determined variables of

> sex that hermaphroditic data had shown could be independent of one another, namely, chromosomal sex, gonadal sex, internal and external morphologic sex, and hormonal sex (prenatal and pubertal), to which was added a sixth postnatal determinant, the sex of assignment and rearing. . . . The seventh place at the end of this list was *an unnamed blank that craved a name*. After several burnings of the midnight oil I arrived at the term, gender role, conceptualized jointly as private in imagery and ideation, and public in manifestation and expression. (Money 1995, 21, emphasis added)

The "hermaphroditic data" led Money to the hypothesis that biological sex is itself radically unstable, composed of heterogeneous elements that do not add up to a unitary conceptual entity. Reckoning with this instability produced for Money a problem of language and reference, a problem of naming (earmarked by his peculiar tautology "an unnamed blank"). The "unnamed blank that craved a name" that Money refers to in this passage can be read as a displacement of the biological instability exposed by intersexuality. In other words, in recognizing a list of prenatally and postnatally "determined variables of sex that hermaphroditic data had shown could be independent of one another," Money's research dismantled the unitary conception of sex and, in so doing, produced an "unnamed blank" at the site of the body. This "unnamed blank" threatened the very semblance of sex. To contain that threat, Money filled the blank with gender. Put differently, Money used gender role to name and thereby semantically fill (or cover over) the void left by sex's lack of conceptual and referential unity. As Germon (2009) puts it, "at a pragmatic level, gender provided a solution to the uncertainty of any absolute somatic sex. Gender served to stabilize what advances in medical technology had rendered more and more unstable during the first half of the twentieth century" (25).

While *gender role* offered stability where technology's destabilization of sex was concerned, it also gave Money a linguistic means to contain the discursive proliferation ("too many words") occasioned by his research on intersex. By giving the "unitary name" *gender role* to the "unnamed blank," Money introduced a seemingly coherent sign where he previously had found only unstable, discontinuous elements. Moreover, Money anthropomorphizes the "unnamed blank"—he attributes to it the "craving" for "a name"—making it seem as if the unnamed blank were itself a subject of desire, longing for epistemic certainty and representational unity, yearning, in short, for someone to give it a name. Giving the "unnamed blank" the name *gender role*, Money proceeds as if that naming could guarantee a relation of referential coherence between word and inchoate object. This anthropomorphism dissimulates Money's own medico-scientific craving for epistemic positivity. By figuring *gender role* as the name craved by the unnamed blank, Money thus overrides and conceals intersexuality's undoing of the structure and stability of sexual dimorphism, and makes the internal and external manifestation of masculinity or femininity the pinnacle of his classificatory schema. In this way, Money posited *gender role* as a predictive agent to determine the hermaphrodite's sex. In short, long before Butler, Money proposed that gender precedes sex.

In contemporary feminist theory, the postulation of gender as prior to sex has been a touchstone for antifoundationalist accounts of embodiment. For Butler (1990), for instance, the reversal of the conceptual polarity of the sex/gender distinction represents the first subversive gesture in a two-pronged deconstructive movement of reversal and displacement. But it is crucial to recognize that Money's superordination of gender over sex was not a subversive gesture but rather a regulatory one. By determining a hermaphroditic infant's prospective gender role, Money was then retroactively able to determine the infant's sex as male or female, and this is why his treatment recommendations centered on surgical, hormonal, and psychosocial normalization. In using "gender role" to fill the "unnamed blank" intersexuality represented, Money attempted to make individuals born with intersex characteristics fit into normative schematizations of the roles conventionally embodied by people with dimorphic sex.

As Iain Morland observes in "Cybernetic Sexology," Money claimed to think "cybernetically" about sex and gender, and his usage of the word "variable" in this instance is a prime example (2015). However, according to Morland, Money made a formative error "in his application of cybernetics to sexology. Cybernetics theorized dynamic systems that can adapt, not merely repeat. It was therefore irreconcilable with the sudden, irrevocable establishment of gender in infancy that was axiomatic for Money" (101). At the very moment when his research pointed toward potentially radical instabilities between gender and sex—and within gender and sex themselves—Money erased those possibilities by reducing gender to the performance of the roles he thought dimorphic sex *should* entail; that is, by fixing gender as mere repetition, as axiomatic, to use Morland's terms. [. . .]

As Morland also earlier noted, the role of gender in the development of intersex treatment, and in Money's research in particular, remains contentious (2009). In the paragraph from his 1955 article "Hermaphroditism, Gender, and Precocity in Hyperadrenocorticism" in which the term first appeared, Money theorized *gender role* as pertaining specifically to the way in which behavior cannot be causally linked to biological sex: "Cases of contradiction between gonadal sex and sex of rearing are tabulated . . . together with data on endogenous hormonal sex and gender role. The term gender role is used to signify all those things that person says or does to disclose himself or herself as having the status of boy or man, girl or woman, respectively. It includes, but is not restricted to sexuality in the sense of eroticism" (254). Money then offered the following summary conclusion, which I quote at length:

> Chromosomal, gonadal, hormonal, and assigned sex, each of them interlinked, have all come under review as indices which may be used to predict an hermaphroditic person's gender—his or her outlook, demeanor, and orientation. Of the four, assigned sex stands up as the best indicator. Apparently, a person's gender role as boy or girl, man or woman, is built up cumulatively through the life experiences he [*sic*] encounters and through the life experiences he [*sic*] transacts. Gender role may be likened to a native language. Once ingrained, a person's native language may fall into disuse and be supplanted by another, but it is never entirely eradicated. So also a gender role may be changed or, resembling native bilingualism may be ambiguous, but it may also become so deeply ingrained that not even flagrant contradictions of body functioning and morphology may displace it. (258)

[. . .] In this passage, Money is not only contemplating gender's moldability but also simultaneously prefiguring and effacing one of the lessons of poststructuralist feminisms: that gender is structured like a language, a system of differences without positive terms (Saussure [1916]

1998; Johnson 1987). If gender is like a language, then gender is not only a relational system but also a system where the meaning of any given term is both arbitrarily and negatively determined. But Money forecloses this insight by positing the existence of proper, positive binary terms as the ground of the system: "*his* or *her* outlook, demeanor, and orientation" (Money 1955, 258). Money's normative dimorphic prerogative and his investment in the propriety of binary logics come together to privilege heteronormative masculine and feminine roles and bodies as regulatory ideals, over and above alternative possibilities of comportment, identification, and embodiment.

Money's reference to "native bilingualism" as "ambiguous" is also noteworthy. The figure marks native bilingualism as indefinite, unclear, and confusing, when in fact native bilingualism just means that a person grows up speaking two languages. Bilingualism opens up opportunities for translation, raises questions about linguistic and cultural difference, and reveals the promise of border crossing. It destabilizes those nations and cultural traditions that privilege the idioms of monolingualism and ethnocentrism (Derrida 1998). Money codes categories and bodily configurations that trouble expected boundaries and forms, disrupt cultural norms and preconceptions, and challenge ideas of sovereignty and wholeness as a threat to intelligibility. As with the "unnamed blank" analyzed above, Money's diagnostic effort becomes regulatory, an effort to contain that which generates ambiguities and proliferates languages and meanings.

The regulatory aspect of Money's work is especially apparent in the gendered language that shapes the passage I have been reading. Between the first and third sentence, there is a grammatical shift from *his or her* to *he*. [. . .] Money switches to the masculine singular pronoun, using it as the general form of personhood. This usage reveals the masculinism, or, more precisely, the masculine universalism that guides Money's project, a masculine universalism evident not only at the level of grammar but also in the conceptual transition from hermaphroditism to binary gender. Money resolves the tension between the destabilization and multiplication of sexes and sexed subject positions inaugurated by his research on intersexuality and binary grammar by privileging the masculine singular pronoun as the signifier of universal personhood.

In other words, Money theorized sex as surgically malleable and gender as socially plastic to maintain the binary order of things. As Morland observes, "the notion of human genitals and gender as surgically and socially plastic depends on the conceptualization in twentieth-century science of plasticity as a quintessential human attribute" (2015, 69). Money conceptualized, Morland continues, "genitals and gender as malleable at a historical moment when plasticity and humanity were held by Western science to be equivalent. This had the mutually reinforcing effects of facilitating the uptake of Money's ideas about how to treat intersex, while instituting gender as a core human quality, flexible by definition" (69). That is, Money's use of gender as a predictive agent presumed that humans are plastic enough to tolerate treatment in the first place.

In devising a course of treatment for intersexuality, Money, along with fellow researchers at the Johns Hopkins Psychohormonal Research Unit, formulated what has come to be known as the optimal gender paradigm. They held that "the sex of assignment and rearing is consistently and conspicuously a more reliable prognosticator of a hermaphrodite's gender role and orientation than is the chromosomal sex, the gonadal sex, the hormonal sex, the accessory internal reproductive morphology, or the ambiguous morphology of the external genitalia" (Money, Hampson, and Hampson 1957, 333). As Vernon Rosario explains, the Hopkins team "argued that infants born with ambiguous genitalia could be surgically 'corrected' and then successfully raised as either males or females so long as certain conditions were met" (2007, 267). These conditions included gender assignment before eighteen to twenty-four months; that parents strictly enforced the gender of rearing; and that the children were "not confused by knowledge about their intersexed past" (267). The optimal gender paradigm "held that *all* sexually ambiguous children should—indeed must—be made into unambiguous-looking boys or girls to ensure unambiguous gender identities" (Dreger and Herndon 2009, 202). In other words, if gender is like language, and gender instability (changing genders) is like native bilingualism, Money's ultimate goal was to eradicate ambiguity in the name of promoting monolingualism. This seems to resolve the problems of both discursive excess ("too many words") and linguistic inadequacy ("an unnamed blank that craved a name").

In recommending that intersex infants be treated with a combination of normalizing genital surgeries, hormonal treatments, and psychosocial rearing into the "optimal gender," Money and his colleagues essentially designed a program of sex *and* gender normalization. This program of normalization can also be understood as a refinement of the masculinism (disguised as grammatical) inherent in Money's privileging of the masculine pronoun. As Karkazis points out, Money and other intersex medical specialists' intentions were, to some degree at least, beneficent: "Raising a child with a gender-atypical anatomy (read as gender ambiguity) is almost universally seen as untenable in North America: anguished parents and physicians have considered it essential to assign the infant definitively as male or female and to minimize any discordance between somatic traits and gender assignment" (2008, 7). Money and the Hopkins team thought that their treatment protocols would help intersex children to live "normal" lives. Intersex activists and scholars have criticized these protocols, however, for inflicting profound physical and psychological trauma and upholding an unjust system of bodily and psychical regulation.

Conclusion

Through Money's work, gender became one of the cornerstones of the modern medical management of intersex. In *Gender Trouble*, Butler observes that "the mark of gender appears to 'qualify' bodies as human bodies; the moment in which an infant becomes humanized is when the question, 'is it a boy or girl?' is answered. Those bodily figures who do not fit into either gender fall outside the human, indeed, constitute the domain of the dehumanized and the abject against which the human itself is constituted" (1990, 111). Seen in this light, Money's project essentially concerned the humanization of people with intersex traits, and unwittingly revealed how dehumanizing humanism can be for those born with anatomies that do not conform to a mythical norm. Though Money's work has been questioned in recent years, many clinicians continue to follow his guidelines, viewing intersex infants as corporeally unintelligible at the moment of birth, only to immediately transport them into intelligibility through surgical, medical, and psychosocial normalization. As my analysis has shown, these bodily interventions follow the strict, masculinist-as-universalizing, binary constraints of a cultural grammar. Most parents and doctors are so overly invested in the question "Is it a boy or girl?" that they cannot imagine a world of other possibilities. [. . .]

This is precisely why a genealogical approach to the messy relations between bodies and the words and practices that name them is so important. A queer feminist science studies approach to the history of intersex disrupts and displaces gender's presumed coherence and meaning; reveals that gender cannot be reduced to a transhistorical given or a purely descriptive category; calls attention to the power relations that transect the lives of people whose bodies have been marked as gender's constitutive outside; and underscores the historical processes, antagonisms, and complicities that have shaped the development of gender as a concept, object of knowledge, paradigm of sociality, and technology of subject formation. [. . .] Intersex literally gave birth to gender.

Works Cited

"INSA webpage." http://isna.org.

Butler, Judith. 1990. *Gender Trouble: Feminism and the Subversion of Identity*: Psychology Press.

———. 1999. "Bodies that Matter." In *Feminist Theory and the Body: A Reader*, edited by Janet Price and Margrit Shildrick, 235–245. Taylor & Francis.

Chase, Cheryl. 2002. "Affronting Reason." In *GenderQueer: Voices from beyond the Sexual Binary*, edited by Joan Nestle, Clare Howell and Riki Anne Wilchins, 297 p. Los Angeles: Alyson Books.

Davis, Georgiann. 2015. *Contesting Intersex: The Dubious Diagnosis, Biopolitics: Medicine, Technoscience, and Health in the 21st century*. New York: NYU Press.

Derrida, Jacques. 1998. *Monolingualism of the Other, or, the Prosthesis of Origin*. Translated by Patrick Mensah, *Cultural Memory in the Present*. Stanford, CA: Stanford University Press.

Downing, Lisa, Iain Morland, and Nikki Sullivan. 2015. *Fuckology: Critical Essays on John Money's Diagnostic Concepts*. Chicago: University of Chicago Press.

Dreger, Alice D, and April M Herndon. 2009. "Progress and Politics in the Intersex Rights Movements: Feminist Theory in Action." *GLQ: A Journal of Lesbian and Gay Studies* 15 (2): 199–224.

Fausto-Sterling, Anne. 2000. *Sexing the Body: Gender Politics and the Construction of Sexuality*. 1st ed. New York: Basic Books.

———. 1993. "The Five Sexes." *The Sciences* 33 (2): 20–24.

Germon, Jennifer. 2009. *Gender: A Genealogy of an Idea*. 1st ed. New York: Palgrave Macmillan.

Hausman, Bernice L. 1995. *Changing Sex: Transsexualism, Technology, and the Idea of Gender*. Durham: Duke University Press.

Hird, Myra J. 2000. "Gender's Nature Intersexuality, Transsexualism and the 'Sex'/'Gender' Binary." *Feminist Theory* 1 (3): 347–364.
Holmes, Morgan. 2008. *Intersex: A Perilous Difference*. Selinsgrove: Susquehanna University Press.
Johnson, Barbara. 1987. *A World of Difference*. Baltimore: Johns Hopkins University Press.
Karkazis, Katrina Alicia. 2008. *Fixing Sex: Intersex, Medical Authority, and Lived Experience*. Durham: Duke University Press.
Kessler, Suzanne J. 1990. "The Medical Construction of Gender: Case Management of Intersexed Infants." *Signs* 16 (1): 3–26.
Meyerowitz, Joanne J. 2004. *How Sex Changed: A History of Transsexuality in the United States*. 1st paperback ed. Cambridge, MA: Harvard University Press.
Money, John. 1952. "Hermaphroditism: An Inquiry into the Nature of a Human Paradox." Ph.D Dissertation, Harvard University.
———. 1955. "Hermaphroditism, Gender and Precocity in Hyperadrenocorticism: Psychologic Findings." *Bulletin of the Johns Hopkins Hospital* 96 (6): 253.
———. 1995. "Lexical History and Cocnstructionist Ideaology of Gender." In *Gendermaps: Social Constructionism, Feminism, and Sexosophical History*, 165 p. New York: Continuum.
Money, John, and Anke A. Ehrhardt. 1972. *Man & Woman, Boy & Girl: The Differentiation and Dimorphism of Gender Identity from Conception to Maturity*. Baltimore: Johns Hopkins University Press.
Money, John, Joan G Hampson, and John L Hampson. 1957. "Imprinting and the Establishment of Gender Role." *AMA Archives of Neurology & Psychiatry* 77 (3): 333–336.
Morland, Iain. 2009. "Introduction: Lessons from the Octopus." *GLQ: A Journal of Lesbian and Gay Studies* 15 (2): 191–197.
———. 2011. "Intersex Treatment and the Promise of Trauma." In *Gender and the Science of Difference: Cultural Politics of Contemporary Science and Medicine*, edited by Jill A. Fisher, vii, 249 p. New Brunswick, NJ: Rutgers University Press.
———. 2015. "Cybernetic Sexology." In *Fuckology: Critical Essays on John Money's Diagnostic Concepts*, edited by Lisa Downing, Iain Morland and Nikki Sullivan, viii, 205 pages. Chicago: University of Chicago Press.
Preves, Sharon E. 2002. "Sexing the Intersexed: An Analysis of Sociocultural Responses to Intersexuality." *Signs* 27 (2): 523–556.
———. 2003. *Intersex and Identity: The Contested Self*. New Brunswick, NJ: Rutgers University Press.
Reis, Elizabeth. 2009. *Bodies in Doubt: An American History of Intersex*. Baltimore: Johns Hopkins University Press.
Rosario, Vernon. 2007. "The History of Aphallia and the Intersexual Challenge to Sex/Gender." In *A Companion to Lesbian, Gay, Bisexual, Transgender, and Queer Studies*, edited by George E. Haggerty and Molly McGarry, 262–281. London: Blackwell.

Saussure, Ferdinand de. [1916] 1998. *Course in General Linguistics*. Translated by Roy Harris. Illinois: Open Court Publishing.
Sullivan, Nikki. 2015. "The Matter of Gender." In *Fuckology: Critical Essays on John Money's Diagnostic Concepts*, edited by Lisa Downing, Iain Morland and Nikki Sullivan, viii, 205 pages. Chicago: University of Chicago Press.

PART TWO

Contemporary Archives and Case Studies

IN THE HISTORY OF WESTERN BIOSCIENCES, THE ARCHIVE AND THE case study are intimately entangled as technologies of knowledge and power (Foucault 2003). Archives and case studies alike serve particular disciplinary purposes. They not only represent but also help to conjure, crystalize, and predicate various subject-formations. Formal conventions of the archive and the case study have been studied from a wide range of interdisciplinary perspectives (Cvetkovich 2003, 2012; Taylor 2003; Stoler 2010; Derrida 1996). Archives are not merely passive repositories from which to extract data, but sites that bear the material traces of practices of governance, biopower, and resistance. Likewise, case studies are technologies of normalization, even as they offer opportunities to think through the specificity of particular questions and problems in relation to more general concepts, epistemic schemas, and context-bound biocultural structures. Contemporary archives and case studies are useful objects of queer feminist inquiry, then, because they expose knotted entanglements of science and embodiment with history, culture, and discourse.

This section collects a series of essays that analyze the production of normality and deviance in a variety of contemporary archives and case studies, including medical cases and archives, pornographic archives, and antiviolence activism. The essays gathered here ask questions about how particular bodies, affects, psychosomatic states, and differences not only are represented, biologized, and naturalized in, but also actively trouble the coherence and stability of a diverse cross-section of archives and case studies. Collectively, they ask: How might queer feminist considerations of contemporary archives and case studies lead to new models of knowledge production, embodiment, and politics?

This question of course builds on a legacy of older feminist and queer questions, such as: How do bodies come to matter? How do subjects become embodied? What tools best allow us to understand the processes and forces that shape the materializations of bodies? How do we talk about pain, pleasure, and other realities of flesh (Spillers 2003), without essentializing bodily properties as evidence in service of naturalizing the status quo? Contributors explore these abiding questions from multiple interdisciplinary angles and in exciting and innovative ways. Read side by side, the essays importantly suggest that the retrenchment of strict disciplinary divides between the hard sciences, social sciences, and humanities forecloses more nuanced understandings of the relationships among history, embodiment, science, and power-knowledge. Investigating diverse intersectionalities, the contributors not only analyze the exclusions and limits of contemporary archives and case studies but also enrich and transform contemporary scientific, feminist, and queer paradigms of study in the process.

In "Black Anality," Jennifer Nash provides a deft analysis of the conflation of blackness and anality and the place of black female flesh in their coformation not in "science" as such, but rather in pornographic discourse surrounding human anatomy and physiology. In addition to citationally foregrounding critical intersections of black feminist thought and science studies (in reproductive justice scholarship and work on the figure of Saartjie Baartman), this excerpted piece, reprinted from a *GLQ* special issue on "The Visceral," does two additional kinds of work important to the project of queer feminist science studies. First, it offers analytics for thinking the physicality of embodiment that depart from and enrich those offered by a notion of the "material" often conflated with science (Irni 2013; Roy and Subramaniam 2016; Willey 2016). Second, Nash's close readings both critique racist and sexist constructions of sexuality and explore humor, parody, and play as generative potentialities in sexual embodiment. In taking the physicality of bodies as a site of resistance, Nash opens up possibilities for new racial/sexual economies.

The essay "At the Same Time, Out of Time: Ashley X" by Alison Kafer comes from her 2013 book, *Feminist, Queer, Crip*. In this piece, Kafer examines the case of Ashley X, a girl with significant cognitive disabilities who was given estrogen to "attenuate" her growth and a hysterectomy and bilateral mastectomy to "reduce the complications of puberty" and mitigate the effects of the estrogen "treatment." Kafer reads the case through the lens of temporality, pointing out that Ashley's parents and doctors represented her condition as one of temporal disruption (her body was developing faster than her mind), justifying medical intervention. At the same time, Kafer demonstrates that many commentators on the case clearly feared Ashley's potential to develop sexuality and reproductive capacity. Thus, Kafer argues, the Ashley case reflects ableism and the erasure of crip desire. This essay represents an innovative piece of feminist, queer, crip scholarship that centers critiques of normative medical logics to imagine crip forms of sensuality.

"'BIID'? Queer (Dis)Orientations and the Phenomenology of 'Home,'" by Nikki Sullivan, brings queer feminist disability studies into these conversations by examining the diagnostic category of "Body Integrity Identity Disorder" (BIID). Through an analysis of the case of Clint Hallam (a man who received a hand transplant and then later had it removed), Sullivan elaborates an understanding of BIID as a "somatechnology" that serves to shore up the boundaries of the normative subject. In addition, by analyzing the way in which the term "walkabout" (a term originally used to oppress indigenous Australians) was applied to Hallam, Sullivan connects the abjection of Hallam (and those diagnosed with BIID) to the abjection of indigenous Australians. The piece represents a shift in focus from those defined as pathological to an examination of the work that illness categories do for those defined as "healthy."

Mitali Thakor's "Allure of Artifice: Deploying a Filipina Avatar in the Digital Porno-Tropics"—an original contribution to this volume—turns to biometric technology and global sex work to offer a theory of the "erotics of artifice." Her intervention into digital community policing is based on ethnographic fieldwork at the United Nations office and antitrafficking organizations in Thailand, child protection NGOs and police headquarters in the Netherlands, and software companies throughout the United States. She analyzes intertwined politics of protection and exploitation—of sexual and savior economies—through a case study, "Project Sweetie," a sting operation designed to catch adult men looking for underage children online using a photorealistic, computer-generated, moving avatar of a ten-year-old Filipina girl. Interrogating how racialized presumptions of vulnerability, desirability, and victimhood are programmed into these efforts, Thakor offers strategies for reading the centrality of racial/sexual formation to global circuits of power that rely upon and reproduce expert authority to say who and what we are and to delimit what types of social justice we can imagine.

Hilary Malatino's "Gone, Missing: Queering and Racializing Absence in Trans & Intersex Archives," also published here for the first time, analyzes the raced and classed absences of intersex and trans archives. Returning to one of the primary architects of trans and intersex medical pathologization, Malatino explores instances of patient disappearance in John Money's case studies. According to Malatino, these instances reveal that access to "technologies of transition" is maldistributed along raced, classed, and gender-nonconforming lines. Through a deeply attentive practice of archival reading, Malatino suggests that sex and gender binarism is reproduced in biomedical discourse and practice through the transit of empire and settler colonialism, to use Jodi A. Byrd's formulation (2011). Malatino frames these absences and deliberate disappearances as a method of resistance to the imposition of what María Lugones (2007) terms the colonial/modern gender system. The essay demonstrates how a queer feminist science studies attuned to intersections of sex/gender, race, and class can foster new ways of understanding the politics of science and genealogies of embodiment.

Discussion Questions

1. What are some of the connections and differences between the archives and case studies analyzed in Part Two? Individually and collectively, how do they help us to understand the normal?
2. What other contemporary archives and case studies would be fruitful objects of queer feminist approaches to reading science?

Works Cited

Byrd, Jodi A. 2011. *The Transit of Empire: Indigenous Critiques of Colonialism*. Minneapolis: University of Minnesota Press.

Cvetkovich, Ann. 2003. *An Archive of Feelings: Trauma, Sexuality, and Lesbian Public Cultures*. Durham: Duke University Press.

———. 2012. *Depression: A Public Feeling*. Durham: Duke University Press.

Derrida, Jacques. 1996. *Archive Fever: A Freudian Impression*. Chicago: University of Chicago Press.

Foucault, Michel. 2003. *Abnormal: Lectures at the Collège de France, 1974–1975*. Translated by Antonella Salomoni and Arnold Ira Davidson. 1st Picador USA ed. New York: Picador.

Irni, Sari. 2013. "The Politics of Materiality: Affective Encounters in a Transdisciplinary Debate." *European Journal of Women's Studies* 20 (4): 347–360.

Kafer, Alison. 2013. *Feminist, Queer, Crip*. Bloomington: Indiana University Press.

Lugones, María. 2007. "Heterosexualism and the Colonial/Modern Gender System." *Hypatia* 22 (1): 186–219.

Roy, Deboleena, and Banu Subramaniam. 2016. "Matter in the Shadows: Feminist New Materialism and the Practices of Colonialism." In *Mattering: Feminism, Science, and Materialism*, edited by Victoria Pitts-Taylor, 23–42. New York: NYU Press.

Spillers, Hortense J. 2003. *Black, White, and in Color: Essays on American Literature and Culture*. Chicago: University of Chicago Press.

Stoler, Ann Laura. 2010. *Carnal Knowledge and Imperial Power: Race and the Intimate in Colonial Rule*. Berkeley: University of California Press.

Taylor, Diana. 2003. *The Archive and the Repertoire: Performing Cultural Memory in the Americas*. Durham: Duke University Press.

Willey, Angela. 2016. "A World of Materialisms: Postcolonial Feminist Science Studies and the New Natural." *Science, Technology, & Human Values* 41 (6): 991–1014.

Black Anality

JENNIFER C. NASH

IN HER FOUNDATIONAL ARTICLE "BLACK (W)HOLES AND THE GEOMetry of Black Female Sexuality," Evelynn Hammonds analogizes black women's sexuality to a "black hole," a space that appears empty but is actually "dense and full" (1994, 138). If black female sexuality is a complex and "full" site, it requires critical practices that "make visible the distorting and productive effects these sexualities produce in relation to more visible sexualities," and that insist on black female *w*holeness (139). Hammonds's work also references another meaning of "black hole," the specter of black female genitalia, the fictive space that has long marked black female sexuality as deviant. How, she asks, does black female sexuality become imagined as a site of difference, as *the* site of difference, and how does the "black hole" become the location of imagined difference?

This essay attends to the "other" black hole—the one that is both overdetermining and undertheorized—and asks how black sexualities generally, and black female sexualities particularly, become tethered imaginatively, discursively, and representationally to the anus. In other words, I ask how the black female anus acts as a significant space through which black sexual difference—and blackness more generally—is both imagined and represented. In this essay, I develop the term *black anality* to describe how black pleasures are imagined to be peculiarly and particularly oriented toward the anus, and thus as peculiarly and particularly attached to *anal ideologies* including *spatiality*, *waste*, *toxicity*, and *filth*.

Reprinted and abridged with permission from Jennifer Nash, "Black Anality," in *GLQ: A Journal of Lesbian and Gay Studies*, Volume 20, no. 4, pp. 439–460.

While I am interested in the production and circulation of anal ideologies and how these ideologies limit black sexual freedom, I also examine moments where black female bodies can attach themselves to these ideologies in ways that engender delight in blackness itself. That is, this essay is not simply an exposé that reveals how anal ideologies constrain and violate black female bodies. Instead, I examine how black women can strategically deploy anal ideologies to expose the kinds of pleasures black subjects can take *in* blackness—its hyperboles and painful fictions—or to expose anal ideologies themselves, revealing how blackness is constructed and produced alongside (and inside) the anal opening.[1]

My interest in examining the centrality of the anus to conceptions of black female sexual difference, and to conceptions of blackness more generally, both builds on and departs from black feminist scholarship. Black feminist theory has long argued that the buttocks are *the* location of imagined black sexual difference: black women's sexual excess is thought to be located in their spectacular buttocks, and black male desire for the buttocks is often taken as evidence of what Aliyyah Abdur-Rahman calls a "cumulative and widespread racial and cultural retardation" (2012, 14). This interdisciplinary body of scholarship is indebted to Sander Gilman's "Black Bodies, White Bodies: Toward an Iconography of Female Sexuality in Late Nineteenth-Century Art, Medicine, and Literature." Gilman (1985) argues that Saartjie Baartman, the so-called Hottentot Venus, became the preeminent symbol of racial-sexual difference in the nineteenth century and that her buttocks were imagined as *the* loci of racial difference (216).[2] He writes, "When the Victorians saw the female black, they saw her in terms of her buttocks and saw represented by the buttocks all the anomalies of her genitalia. . . . Female sexuality is linked to the image of the buttocks, and the quintessential buttocks are those of the Hottentot" (219). Gilman reveals the metonymic work of Baartman's buttocks: Baartman's buttocks represented Baartman, the buttocks represented excessive black sexuality, and Baartman represented all black women.[3]

Baartman's body, which Gilman revealed to be a "master text" of difference in the nineteenth century, has again become a "master text," one that black feminists use to theorize the continued violence that the dominant visual field inflicts on black female flesh.[4] Black feminists—in varied ways—have drawn on Gilman's work to dispel the notion that the buttocks are merely a neutral body part and to show that, as Patricia Hill

Collins argues, "a simple Google search of the term *booty* should dispel doubts—many of the websites clearly link Blackness, sexuality, and African American women" (2005, 151). These scholars emphasize that the buttocks over-determine conceptions of black female sexuality and that the butt remains, as Deborah E. McDowell notes, "the most synecdochical signature of the 'black female' form" (2001, 306). Moreover, scholars including Collins, Janell Hobson, Lisa Collins, T. Denean Sharpley-Whiting, and Deborah Willis have all, albeit in very different ways, traced contemporary cultural preoccupations with black women's buttocks to "a history of enslavement, colonial conquest and ethnographic exhibition—[which] variously labeled the black female body 'grotesque,' 'strange,' 'unfeminine,' 'lascivious,' and 'obscene'" (Hobson 2003, 87). Scholars regularly draw connections between Baartman and other black women whose buttocks have functioned as objects of cultural fascination, including Josephine Baker (Collins 2002, 112), Lil' Kim (Fleetwood 2011), and more recently, Jennifer Lopez (who, despite being Latina, is interpreted through the lens of the Hottentot Venus, (Barrera 2002, 407–16).[5] This body of scholarship has produced certain analytics that now predominate in the study of black female sexualities: *spectacularity* (Scott 2012; Lewis 2012; Davis 2010), *excess* (Fleetwood 2011), *grotesquerie* (Hobson 2003), and *display*.

Even as the buttocks act as an analytic centerpiece of black feminist theorizing on sexuality, visual culture, and sexual politics, many black feminist scholars suggest that the complex cultural meanings attached to the black female buttocks require "further investigation" and remain understudied (Hobson 2003, 88). Abdur-Rahman notes, "There is a dearth of research on the particular significance of black women's asses in popular media culture. . . . Very little attention is paid to the implication of this ass-centricity on wider conceptions of black sexuality and identity, including for black men who are pathologized for desiring black women's purported excessive asses" (2012, 161n30). Similarly, Hobson notes, "the meaning assigned to this aspect [the buttocks] of the black female body has a long and complex history, a history worthy of further investigation" (2003, 88). Yet I argue that rather than focus more attention on the buttocks—attention that often serves only to reify the buttocks as a material site of difference—black feminist theory can productively shift *inward* toward the black female anus, interrogating the meanings generated in and through this space, and exploring how black sexualities are imaginatively and representatively linked to the anal opening.

Indeed, considering black anality—and how "black" and "anal" are often rendered synonymous on black female flesh—opens up a set of new analytics for black feminist theory: *waste*, *toxicity*, and *filth*. Moreover, considering the *spatiality* of the anus allows black feminist scholars to consider how black female sexuality is imagined to be rooted in (and perhaps generative of) certain kinds of filthy spaces, particularly the ghetto. Importantly, these new analytics open up ways to understand black sexualities that prevailing analytics focused on excess cannot fully theorize: how black sexuality, for example, is imagined as dirty (and here, I mean not just metaphorically but literally); how black sexuality is posited as a formation akin to the ghetto: toxic, filthy, and non-reproductive; and how black sexuality is imagined as wasteful. Moreover, these new analytics produce a shift away from a black feminist preoccupation with the visual register, a location that has been imagined to function as the preeminent site of violence, as the locus of the incessant tethering of black female bodies to conceptions of excess.

To study the conflation of "black" and "anal," the unabridged version of this essay performs a close reading of contemporary digital black pornographies, a site where the language of "black holes" [. . .] is omnipresent, and where "black" and "anal" shore each other up both discursively and representationally. [. . .] The essay is not a comparative treatment of how anal pleasures and perils are represented. Instead of presuming that pornography treats black female bodies—or anuses—*worse*, a term that Angela Harris critiques, I ask about the specific strategies that black pornographies deploy to shore up the synonymity of black and anal, and suggest that tracing these strategies is crucial for understanding how "black" and "anal" are linked outside the pornographic lexicon, and for understanding how conceptions of blackness are produced and reproduced in and through black female anuses.[6]

If my work pushes the borders of black feminist scholarship, it also engages queer theory, which has, in varied ways, emphasized the queerness of blackness (e.g. Ferguson 2004, 2007; Johnson 2003; Scott 2010). While I draw on an interdisciplinary body of work that has marked the queerness of blackness, I depart from that tradition, instead drawing on

Abdur-Rahman's insight that "notions of an ass-centered or generally anal sexuality haunt *even heterosexual desiring and coupling* between black people" (2012, 14, emphasis mine). I read the construction of black anality alongside the spectacular discursive heterosexualization of anal sex while considering the role that racialized pornography has played in both shoring up the heterosexualization of black sexualities *and* unsettling it.[7] [. . .]

This essay unfolds in two parts: first, I place my work in conversation with other anal theorists, showing how my investment in black anality both builds on and departs from existing scholarship. The second part of the essay mobilizes the theoretical framework I develop to perform close readings of images from the digital black pornographic archive. My close readings focus on two representational modes through which blackness and anality become tethered: the production of anal space as analogous to ghetto space, and the representation of black sex as wasteful. [. . .][8]

Theorizing Black Anality

In constructing a theory of black anality, I use what Gayatri Gopinath terms a "scavenger methodology" (2005, 22): I track the ghosts of black anality that haunt the scholarly archive and forge continuities between theoretical traditions still constructed as separate, namely, black feminist studies and queer theory. In doing so, I strategically pull from the work of several anal theorists—Leo Bersani, Darieck Scott, Richard Fung, and Kathryn Bond Stockton—to heed Abdur-Rahman's plea that we develop a rigorous understanding of the mechanisms through which black sexuality is posited as "generally anal" (2012, 14). My work, though, is not simply a weaving together of the strands of various scholarship; instead, in this section of the essay, I draw on this varied interdisciplinary scholarship to fashion a theory that can analyze how "black" and "anal" are imagined as synonymous representationally, discursively, and ideologically. It is this theory—one attentive to the particular ways that "black," "female," and "anus" align—that informs the close readings in the essay's second half.

Bersani's "Is the Rectum a Grave?" perhaps one of the most canonical pieces on anality, both examines the profound stigmatization of gay men's imagined enjoyment of anal sex and reveals the potentially redemptive aspects of the anal opening. Bersani argues that the pathologizing of gay male anal sex is analogous to the pathologizing of prostitution and asserts that "the public discourse about homosexuals since the AIDS crisis began has a startling resemblance . . . to the representation of female prostitutes in the 19th century 'as contaminated vessels, conveyancing "female" venereal diseases to "innocent" men'" (1987, 211, internal quote from Watney 1987, 33–34). In both cases, gay men and prostitutes are imagined to "spread their legs with an unquenchable appetite for destruction" (211). Gay men, Bersani asserts, are representationally and ideologically feminized, since the state of being penetrated, and the state of being open to penetration, is culturally and ideologically equated with female "powerlessness," with a willingness to "abdicate power" (212). What Bersani shows is that, as Janet Halley (2006) notes, "in misogyny, in anti-gay-male homophobia, *and* in gay male erotic longing, the vagina and the anus are figured as sexually insatiable and as animated erotically by a desire for annihilation" (151).

If Bersani reveals that gay men are culturally feminized, he also positions the (male) anus as an entry point into a kind of productive and redemptive passivity. He writes:

> Gay men's "obsession" with sex, far from being denied, should be celebrated—not because of its communal virtues, not because of its subversive potential for parodies of machismo, not because it offers a model of genuine pluralism to a society that at once celebrates and punishes pluralism, but rather because it never stops re-presenting the internalized phallic male as an infinitely loved object of sacrifice. Male homosexuality advertises the risk of the sexual itself as the risk of self-dismissal, of *losing sight* of the self, and in so doing it proposes and dangerously represents *jouissance* as a mode of ascesis. (1987, 222)

Bersani proposes an anal ethics where the imagined location of gay male shame—the anal opening—is a site of productive and transgressive self-shattering, and where the rectum acts as a grave where ideas of selfhood are productively—and perhaps pleasurably—buried. Of course, the kind of redemptive and promising self-shattering that Bersani traces seems exclusively attached to gay men's anuses, and the possibility that female anuses could also be locations of this transgressive self-shattering remains untheorized.

Richard Fung's "Looking for My Penis: The Eroticized Asian in Gay Video Porn" also turns its attention to the cultural meanings of the anus, yet theorizes pornographic productions of racialized analities. Fung carefully traces the racialization of the "bottom" and argues that on the pornographic screen, "Asian and anus are conflated" such that pornography "privilege[s] the penis while always assigning the Asian the role of the bottom" (2005, 121). In short, pornography feminizes Asian men—here, Fung sounds in the register of Bersani—such that Asian becomes shorthand for "bottom." Where Bersani sees anal penetration as a locus of productive self-shattering, Fung sees painful humiliation, racism, and the production of racialized economies of desire that "secure a consensus about race and desirability that ultimately works to our [Asian/American men's] disadvantage" (124). Indeed, Fung expands his critique of Asian men's imagined femininity beyond pornography, treating sex itself as *both* "a source of pleasure" and "a site of humiliation and pain," and calls for an "independent gay Asian pornography," one that would contain an expanded and broad conception of the erotic. My work draws on Fung's insights that the anus is racialized terrain, that it is a location in and through which imagined racial differences are produced. Yet unlike Fung, I ask how *women's* anuses can become sites of racial-sexual meaning making, and examine how "black" and "anus" get conflated and rendered synonymous producing economies of desirability and degradation.

Like Bersani, Scott examines the productive and potentially pleasurable location of the bottom. For Scott, though, "bottom" refers to "*both* the nadir of a hierarchy (a political position possibly abject) and . . . a sexual position: the one involving coercion and historical and present realities of conquest, enslavement, domination, cruelty, torture, and so on" (Scott 2010, 257). Bottoming, then, is not simply about an anus that is penetrated but also about social subordination, and quotidian and spectacular acts of racial violence. For Scott, the "bottom" is a site of black pain, but it is also a locus of a "counterintuitive black power," and complex black pleasures (2010, 259).[9] His work carefully traces uncomfortable moments when black men take pleasure in bottomness—in both sexual and racial humiliation and degradation—and locate power in abjection.

For Scott, these bottom positions—what he terms the "special intimacy of blackness with abjection, humiliation, defeat"—are deeply gendered (270). Indeed, the "counterintuitive" black power that his book exposes seems to be one that black *men* can put to use. He notes the centrality of men to his study:

> The abject as a mode of working with blackness need *not* necessarily privilege masculinity, vexed or otherwise, nor need it center male actors, subjects, or characters—though this study does both. It does both because it originates in a conversation with work in the fields of gay male and black queer studies, and with the study of black masculinity having its origins in black feminist critiques of masculinism, and also because of the usual essentially arbitrary limitations on project conceptualization (an arbitrariness that cannot but betray a masculinist tilt on my part, at least with regard to this project). (20)

Though Scott acknowledges that the bottom powers and pleasures he traces *could* be experienced by black women, they remain the largely absent subjects of his book.[10] If *fuckedness* is a location of both pain and redemptive possibility, Scott's book begs the questions: What are the ways that black women's sexualities and subjectivities are tethered to the anal opening, and what are the perils and freedoms that come from that attachment? What are the bottom pleasures, pains, and powers that black women experience? And in what ways is blackness constituted by black women's imagined "intimate" attachment to the anus? My work, then, takes Scott's investment in the messy nexus of power, pleasure, and anal politics and centers the subjects whom his book largely ignores, asking about the distinctive ways that black women's bodies become attached to anal ideologies.

If my close readings are informed by the work of three preeminent scholars of (male) anality who examine the pleasures and pains of bottom locations, they are particularly indebted to Kathryn Bond Stockton's work on the "switchpoints" among black, queer, and anal, and her investment in carefully tracing black anal sexual economies and examining the construction of black anal spaces. Stockton's project *Beautiful Bottom, Beautiful Shame* asks about the value of debasement and examines "why certain forms of shame are embraced by blacks and queers, and also black queers, in forceful ways" (2006, 2). My work is particularly indebted to her reading of Toni Morrison's *Sula*, which treats Morrison as an anal theorist (effectively placing Morrison in conversation with Freud) who draws connections between "the bias against queer anality (and against its pleasures)" and the "stigma of people who live at the bottom of an economic scale" (68). Stockton's point of departure is an analysis of the name of the neighborhood that is at the centerpiece of *Sula*: the Bottom. For Stockton, the Bottom links the imagined filth of both queer sex and black social and economic marginality. In short, the Bottom offers a theory of anality, one where the anus becomes a lens through which one can read "major patterns in black history, black labor history, black folks' migrations, signs of black gender, and the tender matter of racial castration," and one that links black space, black sexualities, racial and economic marginality, and queer pleasures (68).

Like Stockton, I use the anus as a window through which one can theorize race-making and racial-sexual economies, and through which one can imagine and theorize the construction of racially marked space. I draw on her interest in the Bottom—or the anus—as space, as queer pleasure, as site of shame, and as locus of surprising power, yet my interest is in how a particular archive (digital black pornographies) *produces* the "switchpoints" that Stockton describes, how it representationally tethers "black" and "anal." In short, I ask how black pornographies represent the anus—the Bottom—as a passageway through which black pleasures, perils, and pathologies are made visible. Unlike Stockton, though, I am invested in treating race and sexuality as co-constitutive categories; while Stockton positions race and sexuality as two discrete structures that meet in historically contextual and dense "switchpoints," my investment in black anality follows Scott's call to "consider how the history that produces blackness is a sexual history" (Scott 2010, 8). In short, my investment in black feminist theory, particularly intersectionality, leads me to read black anality as a site where race and sexuality are made and represented together. [. . .][11]

Pleasures and Perils

This essay has tracked two particular ways that black anality is (re)produced: through analogizing black women's anuses to other racially marked spaces, particularly the ghetto, and through constructing black anuses as emblematic of black sexualities' wastefulness. Yet these are only two of the ways that black anality is articulated, amplified, and represented. In tracing these two representational strategies, my hope is that the analytics that black anality foregrounds might provide new strategies and reading practices for black feminists to consider the production of black sexualities as different, distinctive, pathological, or problematic.

While my interest is in the host of ways that black sexualities are constructed in and through the anus, I am also invested in the ways that the anus—and ideas of black anality—can be a space of play, pleasure, desire, and delight for black subjects. In mapping black anality, my impulse is *not* to offer a critique of yet another set of racial strategies used to police black bodies. Indeed, I think that black feminist work has given us sufficient tools to critique the violence of the visual field and to consider strategies for recovering black female flesh. Rather, I have traced the contours of black anality both to expose another way that blackness is produced in and through sexuality, and to consider the kinds of play, pleasures, and delight that black bodies can take up in the never-ending quest for sexual freedom. If black bodies are tethered representationally and ideologically to the anus, how might we consider making the anus a space that can also please, excite, and arouse, and also a locus where racial stereotype can be playfully performed and unraveled?

Notes

1 I talk more about pleasures in blackness, race-pleasure, in Nash 2014.

2 Zine Magubane's work offers an important critique of black feminist reliance on Gilman's work (2001).

3 Magubane (2001) problematizes black feminist interpretations of Baartman as black.

4 I discuss black feminist engagement with the Hottentot Venus in Nash 2008.

5 Hobson argues, "Dominant culture came to celebrate Lopez's behind as part of a recognition of 'exotic' and 'hot' Latinas, women perceived as 'more sexual' than white women but 'less obscene' than black women. In this way, Lopez's body avoids the specific racial stigma that clings to black women's bodies" (2003, 97).

6 For critiques of the rhetoric of "black women have it worse," see (Harris 1990). The notion that black women are necessarily harmed by representation is something I take up more extensively in Nash 2014.

7 The emphatically heterosexualized discourse of anality circulates *apart* from AIDS (indeed, part of the labor of the heterosexualization of this discourse is to split anality apart from AIDS). That is, the analities that

I study in this essay are representationally produced apart from the discourses of disease, illness, and death. Though I am interested in how black analities are tethered to ideologies of toxicity, the specter of HIV transmission through anality is not something that this essay takes up in detail. See, e.g., Williams 2008, 11–12.

8 *Editors' note:* See the unabridged version for referenced close readings.

9 Importantly, Scott embraces paradox—including the paradoxes of pleasure—in his book. As he notes, "I am going to try to establish in this book that these paradoxes—luxury that is necessity, freedom that is imprisonment, and perhaps surprisingly, their correspondent vice-versa formulations—speak to the very core of what blackness is in our culture and how we embody it" (2010, 3).

10 Scott ends his book by turning attention to Gayl Jones's novel *Corregidora*, and he notes that "there is not a necessary connection between black masculinity or black maleness and abjection, since it is clear that women can be—and by the normative or traditional definitions of gender, often are or supposed to be—'bottoms,' too" (2010, 265).

11 *Editors' note:* See unabridged version for "Spectacular Anuses" and "Wasteful Anuses."

Works Cited

Abdur-Rahman, Aliyyah I. 2012. *Against the Closet: Black Political Longing and the Erotics of Race*. Durham: Duke University Press.

Barrera, Magdalena. 2002. "Hottentot 2000: Jennifer Lopez and Her Butt." In *Sexualities in History: A Reader*, edited by Kim M. Phillips and Barry Reay, 407–16. New York: Routledge.

Bersani, Leo. 1987. "Is the Rectum a Grave?" *October* 43: 197–222.

Collins, Lisa. 2002. "Economies of the Flesh: Representing the Black Female Body in Art." In *Skin Deep, Spirit Strong: The Black Female Body in American Culture*, edited by Kimberly Wallace-Sanders, 99–127. Ann Arbor: University of Michigan Press.

Davis, Carole Boyce. 2010. "Black/Female/Bodies Carnivalized in Spectacle and Space." In *Black Venus, 2010: They Called Her "Hottentot"*, edited by Deborah Willis, 186–98. Philadelphia, Pa.: Temple University Press.

Ferguson, Roderick A. 2004. *Aberrations in Black: Toward a Queer of Color Critique*. Minneapolis: University of Minnesota Press.

———. 2007. "Sissies at the Picnic: The Subjugated Knowledges of a Black Rural Queen." In *Feminist Waves, Feminist Generations: Life Stories from the Academy*, edited by Hokulani K. Aikau, Karla A. Erickson and Jennifer L. Pierce, 188–96. Minneapolis: University of Minnesota Press.

Fleetwood, Nicole R. 2011. *Troubling Vision: Performance, Visuality, and Blackness*. Chicago: University of Chicago Press.

Fung, Richard. 2005. "Looking for my Penis: The Eroticized Asian in Gay Video Porn." *A Companion to Asian American Studies*: 235–253.
Gilman, Sander L. 1985. "Black Bodies, White Bodies: Toward an Iconography of Female Sexuality in Late Nineteenth-Century Art, Medicine, and Literature." *Critical Inquiry* 12 (1): 204–242.
Gopinath, Gayatri. 2005. *Impossible Desires: Queer Diasporas and South Asian Public Cultures*. Durham: Duke University Press.
Halley, Janet E. 2006. *Split Decisions: How and Why to Take a Break from Feminism*. Princeton, NJ: Princeton University Press.
Hammonds, Evelynn. 1994. "Black (W)holes and the Geometry of Black Female Sexuality." *Differences: A Journal of Feminist Cultural Studies* 6 (2–3): 126–146.
Harris, Angela P. 1990. "Race and Essentialism in Feminist Legal Theory." *Stanford law Review*: 581–616.
Hill Collins, Patricia. 2005. *Black Sexual Politics: African Americans, Gender, and the New Racism*. New York: Routledge.
Hobson, Janell. 2003. "The 'Batty' Politic: Toward an Aesthetic of the Black Female Body." *Hypatia* 18 (4): 87–105.
Johnson, E. Patrick. 2003. *Appropriating Blackness: Performance and the Politics of Authenticity*. Durham: Duke University Press.
Lewis, Sydney Fonteyn. 2012. "Looking Forward to the Past: Black Women's Sexual Agency in 'Neo' Cultural Productions." University of Washington.
Magubane, Zine. 2001. "Which Bodies Matter? Feminism, Poststructuralism, Race, and the Curious Theoretical Odyssey of the 'Hottentot Venus.'" *Gender & Society* 15 (6): 816–834.
McDowell, Deborah E. 2001. "Afterword: Recovery Missions: Imaging the Body Ideals." In *Recovering the Black Female Body: Self-Representations by African American Women*, edited by Michael Bennett and Vanessa D. Dickerson, 296–318. New Brunswick, NJ: Rutgers University Press.
Nash, Jennifer C. 2008. "Strange Bedfellows Black Feminism and Antipornography Feminism." *Social Text* 26 (4 97): 51–76.
———. 2014. *The Black Body in Ecstasy: Reading Race, Reading Pornography*. Durham: Duke University Press.
Scott, Darieck. 2010. *Extravagant Abjection: Blackness, Power, and Sexuality in the African American Literary Imagination, Sexual Cultures*. New York: NYU Press.
Scott, Tynisha Shavon. 2012. "Chasing Afrodite: Performing Blackness and 'Excess Flesh' in Film." Master's, University of Texas.
Stockton, Kathryn Bond. 2006. *Beautiful Bottom, Beautiful Shame: Where "Black" Meets "Queer"*. Durham: Duke University Press.
Williams, Linda. 2008. *Screening Sex*. Durham: Duke University Press.

At the Same Time, Out of Time

Ashley X

ALISON KAFER

IN THINKING ABOUT CRIP FUTURITY, I FIND MYSELF HAUNTED BY Ashley X. Born in 1997, the girl known as Ashley X was diagnosed with "static encephalopathy" a few months after her birth. "In the ensuing years," doctors note, "her development never progressed beyond that of an infant," and her doctors held no hope that her cognitive or neurological baseline would improve (Gunther and Diekema 2006, 1014). "At the age of 6 years, she [could] not sit up, ambulate, or use language" (1014). Concerned about their daughter's long-term future, Ashley's parents met with doctors in 2004 to discuss the potential effects of puberty and physical growth on their ability to care for her at home. Together they crafted a two-pronged plan: "attenuate" Ashley's growth by starting her on a high-dose estrogen regimen; and, prior to the estrogen treatment, remove Ashley's uterus and breast buds in order "to reduce the complications of puberty" and mitigate potential side effects of the estrogen treatment (Gunther and Diekema 2006, 1014). According to her parents and doctors, these interventions were necessary for Ashley's future quality of life: they would reduce her pain and discomfort (by removing the possibility of her menstruating or developing breasts) and would enable her parents to continue caring for her at home (by keeping her small enough to turn and lift easily). [. . .]

Abridged from Alison Kafer, "At the Same Time, Out of Time: Ashley X." in *Feminist, Queer, Crip*. (Bloomington: Indiana University Press 2013), 47–68. Reprinted and abridged with permission.

As becomes clear in both parental and medical justifications of the Treatment,[1] the case of Ashley X offers a stark illustration of how disability is often understood as a kind of disruption in the temporal field. Supporters of the Treatment frame Ashley's disability as a kind of temporal disjuncture; not only had she failed to grow and develop "normally," but her mind and body were developing at different speeds from each other. According to this logic, Ashley's body required intervention because her body was growing apart from her mind; physically, her body was developing rapidly, but mentally, her mind was failing to develop at all. As a result, she was embodied asynchrony; her mind and body were out of sync. By arresting the growth of Ashley's body, the Treatment could stop this gap between mind and body from growing any wider. In order to make this argument, Ashley's parents and doctors had to hold her future body—her *imagined* future body—against her, using it as a justification for the Treatment. Without intervention, the asynchrony between mind and body would only grow wider; Ashley's body would become more and more unbearable to her, to her parents, and to those encountering her in public. This future burden, brought on by the future Ashley, could only be avoided by arresting the present Ashley in time. [. . .]

[. . .] [This] chapter reads the case through a temporal framing, focusing on the ways in which Ashley was cast, and cast as, out of time; from the beginning of the case, she has been represented as temporally disjointed, as an eternal child, and as threatened by her future self. In addition, I explore the gendered dimensions and assumptions of the Treatment, detailing how Ashley's femaleness, or future femaleness, rendered her atemporality particularly grotesque. As this story makes painfully clear, not all disability futures are desirable; in other words, the problem is not only the inclusion of disability in our futures but also the nature of that inclusion. I conclude the chapter, then, with a brief reflection about how to imagine desirably disabled futures.

A Case History of the Ashley Treatment[2]

Ashley's surgery took place in July of 2004 at Seattle Children's Hospital and included a hysterectomy, a bilateral mastectomy, and an appendectomy. In January of 2007, the Washington protection and Advocacy system (WpAs) launched an investigation of her treatment, finding that the hysterectomy was conducted in violation of Washington State law, which mandates judicial review prior to the sterilization of patients who do not or cannot consent. The hospital's own ethics committee had recommended judicial review for the hysterectomy, but a lawyer hired by Ashley's parents had sent a letter to Ashley's doctors suggesting that judicial review was not necessary because sterilization was not the main goal of the surgery and Ashley would never have the ability to make child-bearing decisions. After receiving the letter, the doctors had proceeded without judicial review. Seattle Children's Hospital accepted the findings in the WpAs report, agreeing that they had acted inappropriately and agreeing to implement changes to their procedures (Gunther and Diekema 2006; Ashley's Parents 2007; Wilfond et al. 2010; Clarren 2007; Carlson and Dorfman 2007; Marshall 2007).

Documenting the Ashley Treatment

The details of the Ashley Treatment became public almost two and a half years after her surgery. In October 2006, two doctors centrally involved in the case—Dr. Daniel Gunther, a pediatric endocrinologist, and Dr. Douglas Diekema, a pediatric bioethicist—published the results of the growth attenuation therapy in the *Archives of Pediatric and Adolescent Medicine*. Several months later, Ashley's parents launched a blog called *The "Ashley Treatment": Towards a Better Quality of Life for "Pillow Angels."* [. . .]

In their initial article, which focused primarily on the growth attenuation therapy, Gunther and Diekema argue that Ashley will benefit both physically and emotionally from her smaller size. [. . .] Gunther and Diekema frame the growth attenuation therapy as essential to Ashley's future quality of life; without it, they claim, her parents would eventually be unable to care for her at home or to include her in family events.

Gunther and Diekema's article is as interesting for what it excludes as for what it includes. While the WpAs report stressed the hysterectomy, discussing it at length, the two doctors limit discussion of the procedure and its ramifications to a single paragraph. "A word here about hysterectomy is probably appropriate," they concede, casting discussion about the hysterectomy—and, by extension, the hysterectomy itself—as a mere side issue to the more important topic of growth attenuation (Gunther and Diekema 2006, 1015). The hysterectomy is apparently so trivial, or so incidental, as not to merit extensive analysis on its own; they do not even use the word "sterilization" in regard to Ashley, thereby avoiding that conversation altogether. [. . .]

Effectively rendering Ashley's breasts as even more expendable than her uterus, Gunther and Diekema do not mention the bilateral mastectomy at all—nor does Diekema in an interview with CNN a few months later (Burkholder 2007). [. . .]

Ashley's parents, however, understand the mastectomy differently, representing it on their blog as an essential component of "the Ashley Treatment"; for them, the hysterectomy, mastectomy, and estrogen regimen are all of a piece. The mastectomy, or, to use their language, "breast bud removal," was necessary for three reasons.[3] The primary reason for the "removal" was that any breast development was likely to cause Ashley pain and discomfort. Breasts would make lying down unpleasant for Ashley ("large breasts are uncomfortable lying down with a bra and even less comfortable without a bra") and would "impede securing Ashley in her wheelchair, stander, or bath chair, where straps across her chest are needed to support her body weight." Those straps would then compress Ashley's breasts, causing further pain and confusion. Buttressing this rationale for the procedure were two "additional and incidental benefits": the bilateral mastectomy would eliminate the possibility of breast cancer or fibrocystic growth, two conditions present in the family; it would also prevent Ashley from being inappropriately "sexualized." According to Ashley's parents, the mastectomy "posed the biggest challenge to Ashley's doctors, and to the ethics committee," but the parents ultimately convinced them of the benefits of the procedure (Ashley's Parents 2007). [. . .]

Out of Line, Out of Time

Always flat-chested, never menstruating, finished growing: for Ashley's parents, the Treatment was undeniably about arresting Ashley's development so that they might continue to lift and carry her without difficulty. Mention of Ashley's flat chest and hysterectomy, however, suggests that more than weight was at stake in their decision. They were also concerned about the developmental disjuncture taking place as her body, which was developing more typically, grew further away from her mind, which "stopped growing . . . when she was a few months old." They understood

Ashley's body as en route to "adulthood," even though her mind was permanently mired in "childhood," and this disconnect required intervention. Doctors and bioethicists following the case echoed this concern; the Treatment was necessary to keep Ashley's cognitive self and physical self aligned. The Ashley Treatment thus enacted a circular temporal logic: Ashley's disabilities rendered her out of time, asynchronous, because of this developmental gap between mind and body; her development needed to be arrested to correct this mind/body misalignment; this arrested development then cast her further out of time, more befitting her permanent cognitive infancy.

From the beginning, the Treatment was described as a way to correct the disjuncture between Ashley's body and mind. "When you see Ashley," Dr. Diekema tells CNN, "it's like seeing a baby in a much larger body" (Burkholder 2007). Without the Treatment, this disjuncture would only become more pronounced, as Ashley would eventually become not only a baby in a much larger body, but a baby in an *adult's* body. What was needed, as her parents put it, was to bring Ashley's "physical self closer to [her] cognitive self" (Ashley's Parents 2012). As John Jordan argues, "Despite her otherwise healthy prognosis, Ashley's body had to be articulated as 'wrong' in such a way that the Treatment could be recognized as the best way to make her 'right'" (2009, 25). This "wrongness" was framed in terms of a temporal and developmental misalignment between mind and body, "the brain of a 6-month-old" in the body of one much older; to the extent possible, the Treatment corrects that disjuncture (Burkholder 2007). [. . .]

The term "pillow angel" both reflects and perpetuates this linking of disability with infancy and childhood. Ashley's parents explain that they "call her our Pillow Angel since she is so sweet and stays right where we place her—usually on a pillow" (Ashley's Parents 2007). This phrasing paints a picture of infant-like dependency and passivity; it makes it difficult to imagine Ashley as a teenager or a woman-to-be. Thus, much as the estrogen therapy and mastectomy make Ashley look like the permanent child she allegedly is, the "pillow angel" label names her as such. Within this schema, her body, mind, and identity all line up perfectly.

Such alignment is necessary not only to ensure that people treat Ashley "in ways that are more appropriate to [her] developmental age," but also to protect those around her from disruptions in their temporal fields (Gunther and Diekema 2006, 1016). Dr. Norman Fost, a bioethicist who has often written about the case, echoes Diekema's concerns about the problem of mind/ body misalignment:

> [H]aving her size be more appropriate to her developmental level will make her less of a "freak." . . . I have long thought that part of the discomfort we feel in looking at profoundly retarded adults is the aesthetic disconnect between their developmental status and their bodies. There is nothing repulsive about a 2 month old infant, despite its limited cognitive, motor, and social skills. But when the 2 month baby is put into a 20 year old body, the disconnect is jarring. (Mims 2007)

In invoking the image of an adult body with a baby's brain, and assuming such an image prompts repulsion, Fost enters the realm of the grotesque. He positions Ashley as the embodiment of category confusion, of "matter out of place"; the imagined Ashley blurs infancy and adulthood together, troubling cultural understandings of the normative life course. We are to imagine an adult that looks like "us" but can never function or think like us, and this collision of sameness and difference makes us uncomfortable. George Dvorsky, another bioethicist commenting on the case, makes explicit this link to the grotesque. Writing in support of the Treatment, he too praises its ability to "endow her with a body that more closely matches her cognitive state—both in terms of her physical size and bodily functioning." He then goes on to argue that the "estrogen treatment is not what is grotesque here. Rather, it is the prospect of having a full-grown and fertile woman endowed with the mind of a baby" (2006). The disjuncture between mind and body is apparently all the more jarring, all the more *grotesque*, because of Ashley's gender. Within this framework, Ashley's imagined future body is held against her present body and deemed excessive and inappropriate: too tall, too big-breasted, too fertile, too sexual, too *adult* for her true baby nature. The Treatment was thus necessary to prevent this imagined big and breasty body—this grotesque, fertile body—from coming into being. Dvorsky makes clear the unspoken reason why the growth attenuation had to be combined with a hysterectomy; without the latter, Ashley would remain grotesquely fertile. [. . .]

Feminists have long challenged the reduction of women to their reproductive capacities, and the case of Ashley X reveals how disability both complicates and enables that reduction. On the one hand, despite the surgical focus on her reproductive organs, Ashley is understood to be completely removed from the realm of reproduction. What makes the bilateral mastectomy and hysterectomy permissible is the underlying conviction that Ashley will never need or use her breasts and uterus. [. . .] Thus, Ashley's disabilities prevent her from being reduced to her reproductive organs; unlike nondisabled women, she is not to be understood in those terms.

At the same time, however, the Treatment reveals the extent to which the female body is always and only framed as reproductive. Dvorsky's anxieties about Ashley's fertility suggest that disability only renders such fertility more threatening, more in need of containment and intervention. Furthermore, her parents' presentation of her breasts and uterus as irrelevant and unnecessary testifies to the persistence of a reproductive use-value understanding of female bodies. The only purpose of these body parts is reproductive; if reproduction is not in one's future, then these parts can be shed without ethical concern. The centrality of reproductive frameworks to our understanding of what constitutes a woman or a female is what made the mastectomy and hysterectomy possible or imaginable. Ashley's breasts and uterus were never going to serve their real purpose, so they could be dismissed. [. . .]

At first blush, it makes no sense to describe Ashley as cured or the Treatment as a kind of cure for her condition. The Treatment did not improve her cognitive or physical functioning nor was it intended to do so. Yet it is undoubtedly a curative response to disability. Ashley had to be cured of her asynchrony, at least to the fullest extent possible. She also had to be freed of the specter of her future body, the full-sized, large-breasted, menstruating and fertile body to come. Ashley had her imagined body held against her, and held against her in both senses of the phrase: it was this imagined body that justified the Treatment, and it was this imagined body that became grotesque when compared to her present body. [. . .]

The Future Will Be Privatized: The Ashley Case in Context

Discourses surrounding the Ashley Treatment serve as a template not only for future medical interventions or standards of care but also for how to view the place of disability and caregiving in the early twenty-first century. The future invoked by the Ashley Treatment is a wholly privatized one:

disability and disabled people belong in the private sphere, cared for by and within the nuclear family; and the nuclear family should be the sole arbiter of what happens within it. [. . .] We can see traces of this position in the family's insistence that there was no need for judicial review in this case. [. . .]

This rejection of judicial oversight dovetails with long-standing cultural presumptions about the objectivity and authority of Western medicine. [. . .] Dr. Diekema's response to the WpAs recommendations serves as a case in point. Challenging the WpAs demand for the addition of disability advocates to hospital ethics committees, Diekema asserts that "ethics committees are not for people with political agendas" (Ostrom 2007). [. . .]

Thus, parents, with guidance from doctors, are the only ones with standing in such cases. As Ashley's parents explain on their blog, "in our opinion, only parents with special-needs children are in a position to fully relate to this topic. Unless you are living the experience, you are speculating and you have no clue what it is like to be the bedridden child or their caregivers" (Ashley's Parents, 2007). [. . .] Many editorials, commentaries, and blogs personalized and thereby privatized the debate by phrasing it exclusively in terms of familial questions [. . .].

One of the main themes running throughout critiques of the Ashley Treatment is the need for more social support for parents of disabled children. Supporters of the Treatment counter that such services are currently unavailable and that to "abandon" Ashley's parents to "these harsh social and economic realities" would be cruel; "Ashley does not live in a utopian world," Sarah Shannon notes in *Pediatric Nursing*, and to focus on the need for accessible houses or in-home attendant care is a "utopian view of care" (Shannon and Savage 2007, 177). Shannon's read of current realities is unfortunately accurate, but calling any and all talk of social supports as utopian and therefore unreasonable denies the possibility of different futures and different presents. As Adrienne Asch and Anna Stubblefield explain, there are already-existing practices and technologies that make home care easier, such as mechanized lifts that can assist with transfers. Moreover, many "full-size" adults live successfully in independent settings and receive care outside of institutions, even without the kind of growth-stunting interventions that the Treatment involves (Asch and Stubblefield 2010). Completely brushing aside frank talk of social supports renders these kinds of options invisible, such that the Treatment appears as the only real choice parents can make for their children. [. . .]

Unknown Futures, Narrowed Futures: Measuring "Quality of Life"

The Ashley Treatment has been presented as necessary to Ashley's quality of life. Ashley will be "better off" as the result of these interventions, the story goes; her parents and doctors had to intervene in order to protect her from future harms. [. . .] There is no way to know for certain whether the Treatment improved Ashley's quality of life. We have no baseline of "quality" by which to measure, for Ashley or for any of us. Supporters of the Treatment claim medical evidence for their assertion that the Treatment had a positive effect, but they are extrapolating from other cases or other situations. Ashley's parents' long-term quality of life likely improved, given that Ashley will remain easier to lift, and Ashley's quality of life is bound up in her parents'; if they are doing well, the odds are higher that she is doing well. But, again, we cannot know, not for certain, whether the Treatment benefited Ashley's quality of life.

Were the interventions a success in terms of reducing Ashley's pain? I don't know; I can't know. The surgery itself likely resulted in pain both physical and psychological, but perhaps that pain has faded from Ashley's memory. Perhaps that pain, now passed, is less significant than the constant pain of compressed breasts or the recurring pain of menstrual cramps. Or perhaps not. We cannot know the answers to these questions, but they are presented in Treatment-supportive discourses as self-evident. The claim that the Treatment reduced Ashley's pain is taken as fact.

Missing from this discussion of Ashley's quality of life is the possibility of pleasure; how might the Treatment have foreclosed upon a range of potential sites and sources of pleasure? It is possible that Ashley would have developed the large breasts that reportedly run in her family, and it is possible that she would have experienced discomfort from them. It seems equally possible, however, that she would have experienced pleasure from those imagined large breasts: the sensation of her shirt moving against her skin, or of her skin moving against her sheets, or of her own arms brushing against her breasts. Even the tight chest straps holding her in her chair could have been sources of pleasure: perhaps she would enjoy the sensation of support, or take pleasure in the alternation between binding and release as she was moved in and out of her wheelchair. The inability or unwillingness to imagine these pleasures is a manifestation of cultural approaches to female sexuality and disability. It is seemingly inconceivable to imagine Ashley's body—her disabled female body—as the source of any sensation other than pain. We have few tools for recognizing female sexuality, particularly disabled female sexuality, as positive; nor can we recognize the potential for a self-generated and self-directed sexuality.

Ashley's parents see the mastectomy as offering an "additional benefit to Ashley" beyond its elimination of imagined future pain; according to them, the mastectomy will also prevent "sexualization towards [her] caregiver" (Ashley's Parents 2012). Their syntax is odd here. To what does the "towards" refer? Is it meant to imply the possibility of a caregiver taking sexual liberties with Ashley, so that the mastectomy prevents caregivers from sexualizing her? Or does it refer to the possibility that Ashley might feel sexual when touched by her caregiver? In either case, it is a troubling rationale for the surgical removal of her breast buds. A lack of breasts does not render one safe from sexual assault or abuse, and many would argue that such assault is more the result of a desire for power and control than of sexualization. Or, if their concern is more about Ashley feeling sexual (and it is profoundly unclear what they would imagine that to mean, given their positioning of her as a noncommunicative infant), then the surgery has been justified, in part, on the need to diminish Ashley's access to pleasurable sensations. Maybe Ashley experiences pleasure from being held or hugged, from being bathed in warm water or toweled off, from nestling into a fresh bed or feeling the sun on her face. And if we can recognize those physical sensations as human pleasures to which even the disabled are entitled, then why deny her the future possibility of feeling the sensations of her breasts? [. . .]

At the Same Time, Out of Time; or, Looking for Ashley among Crips and Queers

[. . .] To return then to where I started: in thinking about crip futurity, I find myself haunted by Ashley X. Of course, Ashley is not the only one doing the haunting. Ashley's parents suggest that there have been other pillow angels who have undergone the Treatment, and, if so, their stories remain unknown; I am haunted by that unknown. I think also of those disabled children who were altered in more traditional but no less invasive ways, children whose stories have not been seen as worth remembering, let alone preserving or disseminating. Perhaps the interventions in their bodies were considered a matter of course, a part of the standard of care, and therefore not prompting judicial review or public response; or maybe they were children who were seen not as figures in a sentimental narrative but as the inevitable and unremarkable casualties of poverty, violence, and inequality. Perhaps the details of their lives were unable to capture the public imagination in the same way a white pillow angel could. Sentimentality has historically and culturally been linked with white middle-class femininity, and Ashley's representation as a "pillow angel" calls to mind these racialized discourses of domesticity and passivity. As Patricia Williams points out, the "pillow angel" label held sway in public discussions of the case in no small part because of Ashley's race and class. Williams doubts, and with good reason, that "a poor black child would have been so easily romanticized as a 'pillow angel'" (2007, 9). Williams uses the case as a reminder that we are more concerned with the quality of some lives than others (even as the steps ostensibly taken to "ensure" that quality reveal profound ableist and misogynist anxiety). [. . .]

Yet supporters of the Treatment argue that disability activists have no bearing on this case because Ashley is too severely disabled to be considered a disabled person. Ashley's parents, for example, refer to her as "permanently unabled" in order to distinguish her from other disabled people; "unabled" is a "new category" that includes "less than 1% of children with disability" (Ashley's Parents 2012). Although she does not argue for this kind of new terminology, Anita J. Tarzian agrees that it might be a "misnomer" to call Ashley disabled. Both disability rights and people-first or self-advocacy movements are concerned with individuals who "have some level of cognitive capacity," she explains, which means that these movements do not have the tools or the rhetoric to address those with "severe neurological impairments" (Tarzian 2007).

Predominant models of disability studies and activism too often *do* skim over such people, and Ashley's situation is not, and never has been, similar to most of us working in disability studies. How, then, are we to understand the differences between our experiences even as we name us all as disabled?

Or, to move in the other direction, how might such an identification—we are all Ashley X—work to trouble the binaries of functional/nonfunctional, physical/developmental, or moderate/severe disability? [. . .]

I want to caution, then, against viewing Ashley as exceptional or her case as a spectacular anomaly. After all, there remains a very real possibility that growth attenuation (and its attendant surgeries) will be performed on other disabled kids, which means that we cannot dismiss the case as a one-time event. More to the point, Ashley herself is not wholly unlike the other disabled people inhabiting the pages of this [essay] or the movements and scholarship discussed here. To see her differently, to accept the representation of her as "unabled" rather than "disabled," is to accept an ableist logic that positions impairment—if "severe" enough—as inherently depoliticizing; "unability" becomes the category that allows "disability" to separate itself from those bodies/minds that remain in the margins.

We will remain haunted by the Ashley case, in other words, if we refuse to look for her among crips and queers, if we refuse to recognize her as part of our work. How might we imagine futures that hold space and possibility for those who communicate in ways we do not yet recognize as communication, let alone understand? Or futures that make room for diverse, unpredictable, and fundamentally unknowable experiences of pleasure? [. . .] As we intervene in the representation of Ashley as abnormally asynchronous or grotesquely fertile, as we interrupt the depiction of her as developmentally and temporally other, we must take care, as feminist disability scholars and crip theorists, not to write Ashley out of our own desirably disabled futures.

Notes

1 Following Laura Hershey (2008), I capitalize "treatment" to distinguish between the specific set of surgical and medical interventions used on Ashley (what her parents term "the Ashley Treatment") and the more abstract, general notion of "treatment" as any set of practices that attempt to solve a problem. As Hershey explains, referring to the interventions as a "treatment" accepts and perpetuates the notion that Ashley's body was sick or wrong and in need of a cure. (8)

2 Editors' Note: This section is a brief summary for the purposes of abridgement. For the full case history, please see the original essay.

3 The parents never refer to the procedure as a mastectomy, only as "breast bud removal," but the hospital billing report clearly lists it as "bilat simple mastectomy." (System 2007, 3).

Works Cited

Asch, Adrienne, and Anna Stubblefield. 2010. "Growth Attenuation: Good Intentions, Bad Decision." *The American Journal of Bioethics: AJOB* 10: 46–48. doi: 10.1080/15265160903441111.

Ashley's Parents. 2007. The "Ashley Treatment": Towards a Better Quality of Life for "Pillow Angels."

———. 2012. AT Summary.

Burkholder, Amy. 2007. "Ethicist in Ashley Case Answers Questions." Last Modified January 11, 2007.

Carlson, David R., and Deborah A. Dorfman. 2007. Investigative Report Regarding the "Ashley Treatment". Washington Protection & Advocacy System.

Clarren, Rebecca. 2007. "Behind the Pillow Angel." *Salon*, February 9, 2007.

Dvorsky, George. 2006. "Helping Families Care for the Helpless." Last Modified November 6, 2006.

Gunther, Daniel F., and Douglas S. Diekema. 2006. "Attenuating Growth in Children with Profound Developmental Disability: A New Approach to an Old Dilemma." *Archives of Pediatrics & Adolescent Medicine* 160: 1013–1017. doi: 10.1001/archpedi.160.10.1013.

Hershey, Laura. 2007. "Stunting Ashley." *Off Our Backs* 37: 8–11.

Jordan, John W. 2009. "Reshaping the 'Pillow Angel': Plastic Bodies and the Rhetoric of Normal Surgical Solutions." *Quarterly Journal of Speech* 95: 20–42.

Marshall, Jessica. 2007. "Hysterectomy on Disabled US Girl was Illegal." *New Scientist*, May 9.

Mims, Christopher. 2007. "The Pillow Angel Case—Three Bioethicists Weigh In." *Scientific American*, January 5.

Ostrom, Carol M. 2007. "Child's Hysterectomy Illegal, Hospital Agrees." Last Modified 2007-05-09 00:00:00.

Shannon, Sarah E., and Teresa A. Savage. 2007. "The Ashley Treatment." *Pediatric Nursing* 33.

System, Washington Protection and Advocacy. 2007. Exhibit R. "Hospital Billing Report," Investigative Report Regarding the "Ashley Treatment".

Tarzian, Anita J. 2007. "Disability and Slippery Slopes." *Hastings Center Report* 37: c3.

Wilfond, Benjamin S., Paul Steven Miller, Carolyn Korfiatis, Douglas S. Diekema, Denise M. Dudzinski, Sara Goering, and Seattle Growth Attenuation and Ethics Working Group. 2010. "Navigating Growth Attenuation in Children with Profound Disabilities. Children's Interests, Family Decision-Making, and Community Concerns." *The Hastings Center Report* 40: 27–40.

Williams, Patricia. 2007. "Judge Not?" *The Nation*, March 12, 2007.

"BIID"? Queer (Dis)Orientations and the Phenomenology of "Home"

NIKKI SULLIVAN

OVER THE LAST DECADE OR SO THERE HAS BEEN INCREASING INTEREST in the desire for the amputation of a healthy limb or limbs. Such desires, once held to be paraphilic, are now largely taken to be symptomatic of what psychiatrist Michael First (2005) calls Body Integrity Identity Disorder (BIID). As this diagnostic term suggests, the disorder thus named, is characterized primarily by a lack of bodily integrity, of a sense of wholeness. Indeed, those experiencing such desires often describe a feeling of disjunction between the selves they are and the bodies they have, and as a result, BIID is regularly posited as analogous to so-called Gender Identity Disorder (GID). The psychiatric construction of these experiences as "disordered" engenders a number of biopolitical effects that this chapter sets out to challenge. First, the lack of integrity felt by those desiring "nonnormative" morphologies is constituted as "disordered": the implication being that the bodies of these individuals are obviously "whole" and yet for some reason they fail to experience their corporeality as such. Second, wholeness is constituted not only as visibly self-evident, but more particularly, as the natural (and therefore ideal) bodily-state that all those who are not disordered simply have. Third, insofar as integrity is taken as fundamental to humanness, then the aim of medicine is to restore the ideal state that has been lost or comprised or is unable to be experienced as such. In short, then, BIID is constituted as an

---

Abridged form of the chapter Nikki Sullivan, "'BIID?' Queer (Dis)Orientations and the Phenomenology of 'Home.'" Reprinted by permission from *Feminist Phenomenology and Medicine* edited by Kristin Zeiler and Lisa Folkmarson Käll, the State University of New York Press

individual(ized) pathology that has little or nothing to do with one's being-in-the-world.

[. . .] This chapter takes a deconstructive approach to BIID, mapping the phenomenological effects of such nomenclature, while simultaneously problematizing its presumed empirical status. In short, I wish to demonstrate that BIID is less the description of an empirical reality than a biopolitical somatechnology; one that establishes and polices boundaries and borders between "us" and "them," between proper and improper bodies—both individual and social—and evaluates their worthiness in terms that replicate (and naturalize) dominant idea(l)s about bodies, selves, and the relations between them.

In order to queer the somatechnics of abjection at work in and through BIID, I want to reorient debates about the desire(s) for amputation, and other forms of "nonnormative" embodiment, away from the question of integrity, with which such debates are primarily concerned and toward a consideration of "orientation." The reason for this, as I explain in more detail later, is twofold. First, the focus on integrity is almost entirely confined to the figure of the person who suffers and does not include an analysis of those who evaluate the alleged lack of integrity and make clinical decisions—or, in the case of a more general public, moral judgments—on the basis of their perception of the "other" who suffers. Second, and relatedly, this limited focus veils over the fact that all who are involved in any consideration of integrity, or, for that matter, in any intercorporeal encounter, are, as Haraway (2008, 224:3) notes, "consequent on a subject- and object-shaping dance of encounters." Given this, it is the dance that Haraway identifies that is of more interest to me than the notion of integrity per se, since the former is illustrative of the ways in which particular somatechnologies[1] of identity and difference operate such that some desires and morphologies are naturalized whereas others are constituted as abject(ed).

BIID: What's in a Name?

Body Integrity Identity Disorder (BIID) is, as I said, a diagnostic term for what is often described as the incessant and insufferable experience of a lack of "wholeness" or "bodily integrity." Researchers have been keen to determine from whence this experience of lack comes, particularly given that the integrity of a fully limbed and fully functional body is commonly taken to be visibly self-evident. But, as Wim Dekkers, Cor Hoffer, and Jean-Pierre Wils (2005) note, "bodily integrity" is less an empirical fact than "an ambiguous notion" shaped by competing moral points of view. This is perhaps not surprising given that historically the term "integrity" has been used to refer not only to the state of being whole, complete, unimpaired, undiminished, untouched, perfect, and so on, but has also been linked etymologically to moral soundness or rectitude: to "uprightness" and/or "straightness"—a point I return to later in the chapter. What gives BIID (as a diagnostic category) its conceptual coherence is, of course, the unspoken (and unquestioned) assumption that the "ordered" individual, the ("normal") person not suffering from BIID (or some other such pathology) experiences his/her body, his/her "self" as "whole." Bodily integrity, as an idealized state, then, takes on the mantle of "the natural," and thus the morally desirable.

Before we consider how this particular understanding of bodily integrity operates in accounts of the desire for amputation, blindness, deafness, and so on, I want first to very briefly discuss its deployment in debates about other modificatory medical procedures. My aim in doing this is twofold: first, I want to (re)situate the body projects associated with "BIID" in the context of (trans)formative practices more generally, and second, I want to critically examine what Dekkers, Hoffer, and Wils refer to as two different approaches to the question of bodily integrity, both of which rely on the notion of integrity as a natural given and a moral good. These they call the "person-oriented approach" and the "body-oriented approach." [. . .]

[. . .] In short, from a body-oriented perspective, the body takes (moral) precedence over the person who temporarily inhabits it, whereas from the person-oriented perspective, the person takes (moral) precedence over the body he or she owns. What becomes clear here, and indeed, what is apparent in most accounts of bodily integrity and/or its lack, is the (problematic) assumption (and reproduction) of a distinction between the body and the self or mind.

While there is much talk of a return to the body in contemporary Western culture, it is nevertheless the case that in the context of neoliberalism, what Dekkers, Hoffer, and Wils refer to as the "person-oriented approach" to bodily integrity is far more common than the "body-oriented approach."

As Tamsin Wilton notes, for example, in accounts of and responses to transsexualism, "it is whatever *inhabits* the transsexual body that matters. . . . The surgeon act[s] *on* the body to ease the pain of the dys/embodied self 'inside'" (cited in Davy 2011, 52). Similarly, so-called cosmetic procedures are more often than not justified on the basis that they furnish the person, whose body is pre-operatively "wrong" and/or at odds with his/her sense of self with the integrity to which she/he allegedly has a natural right. But despite this commonplace privileging of the person over the body, of consciousness over what is constituted as little more than brute matter, the argument that the person suffering from BIID has a natural right to integrity (in the same way as does the cosmetic surgery recipient and/or the person with "gender dysphoria") has not, to date, resulted in access to surgical procedures that might engender a *restitution ad integrum*, that is, a restoration of intactness.

Elsewhere, I have argued that this asymmetry is largely an effect of the (generative) perception of bodies with less than four "full-length," "fully functioning" limbs, with eyes that do not see, as "disabled" (Sullivan 2005). And in the dominant imaginary "disability" is constituted as abject(ed), as the "zone of uninhabitability," the "site of dreaded identification against which—and by virtue of which—the domain of the subject . . . circumscribe[s] its own claim to autonomy and to life" (Butler 1999, 237). Consequently, the desire for amputation (or for blindness, deafness, etc.) as a *restitution ad integrum* is most often perceived by those who do not experience such desires as a contradiction in terms, symptomatic of madness (psychopathology) and/or badness (perversion), and thus as evidence that the "wannabe"[2] lacks the capacity to make informed, rational decisions about his/her well-being and therefore does not have, or should not be accorded, the right to self-determination. This [. . .] functions, I contend, to abject the wannabe from the domain of the subject and to overcome the disorienting experience of being faced with culturally unintelligible desires and morphologies.

One possible response to this sort of ontological stand-off might be to suggest, as Slatman and Widdershoven (2010) do in their work on organ transplantation, that given the seeming centrality of bodily integrity to individual well-being, we have a moral responsibility to articulate and support differential conceptions of "wholeness" (or lack thereof) that could justify diverse treatment protocols. [. . .] Drawing on the significantly different experiences of two hand-transplant recipients, Clint Hallam and Denis Chatelier, [Slatman and Widdershoven] argue that while for the former (who eventually had the transplanted hand removed) the "transplant violated his bodily integrity instead of restoring it," for Chatelier, who "had lost his experience of embodied wholeness" after both his hands and forearms were blown apart when a firework he was carrying detonated, "the transplant restored it" (2010, 86). [. . .]

On the surface, this claim sounds reasonable enough, but what its authors fail to recognize is that the failure to perceive the other's integrity as such is in fact a structural effect of the myth of integrity (as something we have or do not have) that functions, to borrow a phrase from Sara Ahmed (2006, 121, 137), as a "straightening device." [. . .] Consequently, Slatman and Widdershoven's (2010, 88) assumption that "we can contribute to the making of decisions that will . . . respect [the other's] bodily integrity" results not only in the reaffirmation of integrity as a moral good but, by association, in a failure to critically interrogate the ways in which habituated orientations shape our perspective/perception such that our responses (to the other) are, as I show in my discussion of the Hallam case, at once, less than conscious, profoundly affective, and, for the most part, normalizing.

[. . .] In the following section I elaborate this claim not with explicit reference to the notion of integrity but instead by turning to a phenomenological understanding of orientations. My decision to turn away from the notion of integrity is threefold. First, I am mindful of the fact that simply arguing that integrity is a myth does little or nothing for those whose orientation toward particular abjected morphologies is "blocked" and whose being-in-the-world is thus dominated by the experience of alienation, of suffering, and so on. Second, it seems to me that what is at stake in the dominant responses to Hallam and Chatelier, and to those desiring amputation, deafness, bigger breasts, thinness, longer legs, and so on, is not so much integrity, but rather, the position and the status of the "objects" (the morphologies) toward which such desires are oriented and thus, in turn, the status and position of the orientations themselves and of the subjects thus oriented. Third, while I want to maintain a focus on the suffering experienced by those whose desires for or orientations toward amputation, deafness, and so on, are "blocked" I want, for ethico-political reasons that will become clearer in due course, to shift the focus of scrutiny away from these abjected morphologies and interrogate instead the invisible center, the "here," if you like, from which desires and/ or morphologies arrive or are given (in and through particular historical orientations) as "other."

The Clint Hallam International Surgical Soap Opera[3]

In her book *Queer Phenomenology*, Sara Ahmed states that

> the concept of "orientations" allows us to expose how life gets directed in some ways rather than others, through the very requirement that we follow what is already given to us. For a life to count as a good life, then it must return the debt of its life by taking on the direction promised as a social good. (2006, 21)

Orientations on this model are less natural inclinations than the performative effect of the work of inhabitance or dwelling-with; orientations shape and are shaped by our "bodily horizons" or "sedimented histories," and thus are necessarily morphological. [. . .] In and through this process of inhabit(u)ation certain things (i.e., "objects," ways of thinking, styles of being, bodily experiences, and so on) become available to us, while others are constituted as "a field of unreachable objects" (15), as abject(ed). Thus, as Ahmed explains, "we do not have to consciously exclude those things that are not 'on line.' The direction we take excludes things for us" (15)—a point Slatman and Widdershoven appear to forget.

With these insights in mind, let us turn to the figure of Clint Hallam. But let me first stress that this turn to "Hallam" (as a discursive figure shaped by a range of somatechnologies that I discuss in due course) should not be read as suggesting that Clint Hallam ("the man") suffered from BIID. Rather, I am interested in the ways in which the somatechnologies at play across this figure—and in particular, Hallam's decision to have the grafted hand amputated—orient and are oriented by particular idea(l)s about bodies and bodily practices.

On the 13th of September 1998, Hallam, a New Zealander who had lost his right hand in a circular saw accident fourteen years earlier, became the first recipient of a human hand transplant. [. . .] In the popular imaginary the hand transplant signaled the arrival of a time and place in which "anatomical incompleteness" need no longer exist, of a golden age in which all those unfortunate enough to suffer this particular "disability" might be remade whole. [. . .] But this vision of an able-bodied futurity was sorely shaken when, in February 2001, the hand that had held out so much promise, was amputated. However, as both the surgeons involved and the media were quick to point out, the amputation was by no means the result of a failure on medicine's behalf (Campbell 2004, 2000a, 2000b).

In the months following the amputation the international press was awash with stories that, in their attempt to render intelligible Hallam's rejection of "the hand," consistently painted a picture of the New Zealander as a disturbed and disturbing individual, a person whose body had been made "right" but whose being was entirely "wrong" (2000a, Kaebnick 2000). Interestingly, Hallam—who the medical team (allegedly) discovered after the transplant, had spent time in a New Zealand prison for fraud and had seemingly continued to be involved in criminal activities after the operation—is frequently referred to in newspaper reports as "a mercurial character" (2000a), a "trickster" (Campbell 2004, 453), a liar (2000a): as someone who cannot be pinned down, who refuses to stay in (his) place (2000a). [. . .] As Professor Nadey Hakim, one of the doctors on Hallam's team, reported,

> Last time [I] spoke to him [he] was calling from Las Vegas. . . . [I] begged him to look after the hand, to take his medication, to travel—as he was supposed to—to a university in Chicago that had promised to pay him in exchange for being able to perform experiments on him. He [Hallam] said, "Yeah, yeah, I'm in charge of my arm" [says Hakim]. What does he mean? I don't know. It wasn't the ideal choice of patient. (2000a)

Later in the article Hakim returns to the latter point, lamenting the fact that the transplant team had not chosen a patient more like Denis Chatelier for the legendary "first hand transplant" since, claims Hakim, "He [Chatelier] is a decent man. He stays in. He listens to his doctors' advice. He takes his medicine" (2000a).

What most interests me about the vision of Hallam presented by (some) members of his surgical team is what it tells us about the investment in fixing, in stasis, in clear-cut and unchanging boundaries that informs not only their perception of the "troublesome" New Zealander but also their bodily horizon(s), the place(s) from "which they dwell, and the(ir) field of unreachable objects. "We gave him the chance of a lifetime," one surgeon told *The Times*, "and he ruined it . . . he was such a bad example" (Kaebnick 2000).

From this privileged and institutionally authorized perspective, then, the surgeons (who could be said to metonymically stand in for "medicine") gave Hallam a world in which to dwell, a domain in which to be subject; they directed him toward a futurity known (to them) as "the good life," and yet he failed to follow their directions and thereby to reinscribe the familiar path that those coming after him could, in turn, follow. Drawing on Ahmed's work I want to suggest that selecting Hallam (as the candidate for the first hand transplant) and giving him a world in the shape of a hand—a hand that interestingly did not "fit"—constituted an "act of recruitment." Recruitment, writes Ahmed, restores the body of the institution, which depends on gathering bodies to cohere as a body. Becoming a "part" of an institution, which we can consider as the demand to share in it, or even have a share of it, hence requires not only that we inhabit its buildings, but also that we follow its line (Ahmed 2006, 133–34), that we become aligned with it. And of course, Hallam did neither, instead eschewing the "straightening device[s] that function to *hold things in place*" (66).

Let me elaborate on this by turning to a comment made by the Australian surgeon Professor Earl Owen who co-led the team that performed the transplant, and later removed the hand. In response to the interviewer's question about Hallam's alleged failure to cooperate with doctors, Owen states,

> The first three months after the operation he was a good boy, if you like. He was a normal patient, he did his physiotherapy, he had his blood tests, he stayed close to the hospital, which he'd contracted to do, and he was fine, and then he suddenly . . . went walkabout—that's an Australian expression—but it means he disappeared and he set the pattern of disappearing from then on. (2001)

As a fellow Australian I was struck by Owen's use of the colloquialism "walkabout" in this statement. "Walkabout," a term coined by early settlers to describe a little-understood "nomadic" practice (or set of practices) allegedly participated in by (some) indigenous Australians, is, I want to suggest, the product of a white Australian colonialist optics rather than an empirical reality. In the narratives of cultural difference that have, for the last two hundred years or so, circumscribed not only social relations in Australia but also the materialities of those who take up the *socius* and those who are condemned to its margins, to the zone of uninhabitability, "walkabout" has played a significant role.

In an article entitled "The Walkabout Gene" (1976), American anthropologist Charlotte Epstein, who spent time in outback south-central Australia in the early 1970s, relates a series of disturbing encounters that poignantly illustrate this claim. In a discussion about academic achievement with a second-grade (nonindigenous) teacher in a school whose population was 20 percent Aboriginal or "mixed-race," the teacher assured Epstein that "the Aboriginal children could never equal the European children, and that the reason for this was a basic genetic difference that manifested itself in walkabout behavior. [. . .]" (141). Another (white) teacher, from a different school, told Epstein, "The Aborigine is very complicated." And he continued, "You see . . . he can't discipline himself the way we do. He does what needs doing today, but he won't do for tomorrow. And every few weeks, he goes walkabout" (144).

In each of the encounters recounted here, a particular point of view—what we might think of as the "here" of "whiteness"—is taken as given, and from this "givenness" "the Aborigine" and/or the other who cannot inhabit whiteness (or take up the domain of the subject) acquires both a direction and an identity-abject and abjected to the zone of uninhabitability, or, if you like, "homelessness." [. . .] Owen's generative vision of walkabout puts the "nonwhite" other[4] in his place and functions to hold him there by fixing his difference, shaping what he can and cannot do. At the same time, the contours of "inhabitable space" are reinforced in and through this (coincident)[5] encounter such that Owen, his colleagues, and the institution of medicine are once again able to "feel at home," to be positioned and to take up the position of master of all that they survey.

Reconfiguring the Phenomenology of Home

[. . .] As we know, worlds unfold predominantly along (already given) lines of privilege that are the effect of sedimented histories. And, as Ahmed notes, and the outraged directed at Hallam shows, "following such lines is 'returned' by reward, status and recognition" whereas not following them, or not being able to follow them because their mode of operation, their "zero-point of orientation," necessarily excludes the body one is/has from inhabitable space, constitutes the lived experience of some modes of bodily being as "out of place" (Ahmed 2006, 183). [. . .] Again, this claim is poignantly (and, I think, painfully) illustrated by another encounter Epstein relays in "The Walkabout Gene."

In a third school at which Epstein carried out her research, a fourth-grade teacher spoke proudly to the author of how the students (who were about 50 percent white and 50 percent indigenous) got along "Just beautifully. . . . They see each other as all the same—no differences." When Epstein asked the teacher to explain what she meant by this, the teacher responded by showing Epstein some self-portraits the children had recently produced, all of which depicted white children with light-colored hair. Somewhat perplexed, Epstein asked the teacher if any of the children ever portrayed themselves as "dark skinned," to which the teacher replied "No" in a manner that suggested her to be

> comfortable in the rightness of things. Then a slight frown creased . . . [the teacher's] forehead and she leaned toward [Epstein] . . . lowered her voice, [and said] "Do you see that child there?" . . . indicating a very dark [skinned] Aboriginal child. "The other day, when I asked the children to draw themselves, he colored himself very dark. I asked him why, he'd never done that before. And he said to me I'm black, and I'm beautiful." (1976, 144)

I realized, continued the teacher, "that someone must have said something to him and that he was upset. But we straightened it out, and he's fine now." "He draws himself, white-skinned now?" asked Epstein (who had the recently produced "white" self-portraits in front of her). "Yes," answered the teacher. "He's quite alright now. It was just a temporary thing" (144).

There is so much that could be said about this anecdote, but for reasons of brevity I confine my comments to one aspect of the situation, that is, the "seeing queerly" and the "straightening out" of the indigenous child as black *and* beautiful. In her discussion of the spatial experiments recounted by Merleau-Ponty in *Phenomenology of Perception*, Ahmed argues that "queer effects," that is, effects that disorient—such as, for example, when the "here" of whiteness is faced with a vision of the "non-white" body as beautiful, as inhabitable, as the site of an "I"—are overcome through the realignment of what Nirmal Puwar refers to as "matter out of place" (2004, 10), with the bodily horizon. This (re)alignment with lines of privilege depends, as Ahmed (2006, 66) notes, "on straightening devices that keep things in line, in part by 'holding' things in place." Given this, one might argue that the threat posed by the queer vision of the morphological other, the dark-skinned child, is overcome by the "straightening out" to which the teacher refers, that is, the realignment of the domain of the subject with whiteness, the putting of the indigenous other back in his place (i.e., the zone of uninhabitability), and the holding in place of the contours of difference by the repetition of privileged lines (i.e., the indigenous child once again depicts himself as white thus aligning himself with the dominant and naturalized ideals of the "here" of whiteness, which, ironically, he will never be able to inhabit as "home"). [. . .] In much the same way that the idea that black is beautiful or that nonwhite morphologies might be habitable and therefore desirable seemed culturally unintelligible in the context of early 1970s white Australia, encounters with bodies without the so-called full complement of limbs, with hands that do not fit, with ears that do not hear, and so on, could in our current context, be said to produce queer effects, to disorient the lines of privilege that constitute certain morphological futures as necessarily excluded and to make visible lines that disappear from view at the point at which "the subject" emerges (15) and from which that subject apprehends a world. If, then, as Ahmed claims, the question of orientation is about how we "find our way" and how we come to "feel at home," the question of disorientation is about how the encounter with the "unhomely" (the uncanny as the repudiated foundation of the subject's dwelling) queers or denaturalizes habituated modes of dwelling—with, such that "being-at-home" is recast as the contingent effect of ongoing labor that is never mine alone. Of course, as we have seen in the examples discussed throughout this chapter, queer disorientations can be, and often are, divested of their disruptive potential in and through the deployment of a range of straightening devices, one of which is pathologization. In perceiving, and thus naming, the diverse desires and morphologies currently associated with BIID as "disordered," both wrongness and suffering become firmly located in the individual in need of fixing, of straightening out, of realignment. What is denied in this process is the fact that for (many) wannabes the source of suffering lies not in the bodies they want but do not have, nor even in the fact that they desire amputation, deafness, and so on. Rather, suffering is engendered as an effect of a life lived "out of place," of not being at-home-in-the-world or in the body that gives one a world, a "here" from which to extend into phenomenal space and by which to shape that space. Given this, the challenge, I want to suggest, is to move away from the moral imperative to understand and/or respect the others' desires, their morphological difference, and instead to articulate an ethics of dwelling, a critical interrogation of "lines of privilege" (Ahmed 2006, 183), those naturalized positions/perspectives from which particular worlds unfold while others are abjected: "to trace the lines for a different genealogy, one that would embrace the failure to inherit the family [or familiar] line as the condition of possibility for another way of dwelling in the world" (178), one that would admit or even embrace the inherent liminality of "home," multiple evocations of home[6] as always in the making, always (un)becoming—with, as something other, and something more, than the exclusory effect of unexamined inhabit(u)ation. Perhaps, after all, home is not a singular dwelling in which integrity naturally reigns.

Notes

1 The term "somatechnics" is derived from the Greek σώμα (body) and τέχνη (craftsmanship). It is intended to highlight the inextricability of bodily being and the techniques (*dispositifs* and "hard technologies") in and through which morphologies are (trans)formed (Murray and Sullivan 2012).
2 This is a term that some people desiring the amputation of a healthy limb or limbs use to refer to themselves. [. . .]
3 I borrow this phrase from Campbell (Campbell 2004).
4 It is probably worth reiterating here that Hallam is a New Zealander, while Owen is Australian. Historically, New Zealand has been positioned by Australia(ns) as its sort of "poor cousin," its other, if you like. New Zealand is also often represented in the Australian imaginary as marginal to Australia, not only geographically, but in a whole range of ways. In making this claim and/or discussing the incidents related by Epstein it is not my

intention to conflate the position of "Hallam" and that of "indigenous Australians," or to suggest that the mechanisms of abjection operate in exactly the same way in both cases. Rather, my aim is to bring to light some of the somatechnologies of identity and difference that simultaneously naturalize some morphologies and desires, some ways of knowing, seeing, and being and constitute others as abject(ed).

5 I use this term in the Husserlian sense.

6 I borrow this phrase from Fortier (2001).

Works Cited

2000a. "Sleight of Hand." *The Guardian*, May 30 2000. Accessed 2016-06-29 15:46:18.

2000b. "Surgeon Agrees to Sever Transplant Hand." *BBC*, 2000/10/21, Europe. Accessed 2016-06-29 15:48:57.

2001. "Transplanted Hand Removed." Last Modified February 5, 2001.

Ahmed, Sara. 2006. *Queer Phenomenology: Orientations, Objects, Others*: Duke University Press.

Butler, Judith. 1999. "Bodies that Matter." In *Feminist Theory and the Body: A Reader*, edited by Janet Price and Margrit Shildrick, 235–245. Taylor & Francis.

Campbell, Fiona A. Kumari. 2004. "The Case of Clint Hallam's Wayward Hand: Print Media Representations of the 'Uncooperative' Disabled Patient." *Continuum* 18: 443–458. doi: 10.1080/1030431042000256162.

Davy, Zowie. 2011. *Recognizing Transsexuals: Personal, Political and Medicolegal Embodiment*: Ashgate.

Dekkers, Wim, Cor Hoffer, and Jean-Pierre Wils. 2005. "Bodily Integrity and Male and Female Circumcision." *Medicine, Health Care and Philosophy* 8: 179–191. doi: 10.1007/s11019-004-3530-z.

Epstein, Charlotte. 1976. "The Walkabout Gene." *The Urban Review* 9: 141–144. doi: 10.1007/BF02173517.

First, Michael B. 2005. "Desire for Amputation of a Limb: Paraphilia, Psychosis, or a New Type of Identity Disorder." *Psychological Medicine* null: 919–928. doi: 10.1017/S0033291704003320.

Fortier, Anne-Marie. 2001. "'Coming Home' Queer Migrations and Multiple Evocations of Home." *European Journal of Cultural Studies* 4: 405–424. doi: 10.1177/136754940100400403.

Haraway, Donna. 2008. *When Species Meet*. Minneapolis: University of Minnesota Press.

Kaebnick, Gregory E. 2000. "Hand Transplant Recipient Throws in the Towel." *Hastings Center Report* 31: 6–7.

Murray, Samantha, and Nikki Sullivan. 2012. *Somatechnics: Queering the Technologisation of Bodies*. New York: Ashgate.

Puwar, Nirmal. 2004. *Space Invaders: Race, Gender and Bodies out of Place*. New York: Berg.
Slatman, Jenny, and Guy Widdershoven. 2010. "Hand Transplants and Bodily Integrity." *Body & Society* 16: 69–92. doi: 10.1177/1357034X10373406.
Sullivan, Nikki. 2005. "Integrity, Mayhem, and the Question of Self-demand Amputation." *Continuum* 19: 325–333. doi: 10.1080/10304310500176487.

The Allure of Artifice

Deploying a Filipina Avatar in the Digital Porno-Tropics

MITALI THAKOR

Introduction: Project Sweetie

The lights switch on and a child speaks, her face looking directly into the camera. "My name is Sweetie. I'm ten years old. I live in the Philippines," the face on the screen begins in a soft voice.

"Every day I have to sit in front of the webcam and talk to men. Just like tens of thousands of other kids." The video shows a cascade of collaged photos of children, their heads cropped out of the images, as well as photos of adult men holding up panties and other articles of clothing. "The men ask me to take off my clothes. . . . They play with themselves. They want me [to] play with myself. As soon as I go online, they come to me. Ten, hundred, every hour. So many. But what they don't know?"

The camera fades to show neon grid architecture beneath the child's face. "I'm not real. I'm a computer model, built piece by piece." She continues, shaking her head, "To track down these men who do this."

The video continues with a different voice, male and adult, describing how "webcam sex tourism" is a dangerous new form of exploitation, "spreading like an epidemic," as children in places like the Philippines are paid to undress on livestreamed video in chat room websites (Sweetie 2013). The narrator explains that while tens of thousands of children have been forced to perform for "men from rich countries," only six people have ever been "charged." The narrator explains that his team has decided to take matters into their own hands—patrolling the websites where illicit chats with "predators" occur, and posing as a ten-year-old designed to "look and act like a real girl." On the chats, "Sweetie" would try to convince people to share their email addresses, names, and identifying information in order for her to share nude photos and videos of herself. Over the course of ten weeks, the Sweetie team had collected a list of one thousand names and other identifying data of people from seventy-one countries, including the United Kingdom, Australia, Germany, and India. In November 2013, they submitted the list to the Europol office in The Hague and to the Dutch police in South Holland. The narrator of the video concludes by saying that Project Sweetie is an example of the power of technology to draw attention to the webcam sex tourism issue, through which "predators are being stopped, and children are being saved." The team that produced Project Sweetie is named at the end of the video as Terre des Hommes, a Netherlands-based children's charity and nongovernmental organization (NGO). The NGO regards its work as "proactive policing"—with the caveat that no law enforcement officials were involved with the campaign in any capacity.

In this essay, I introduce Project Sweetie as emblematic of new digital techniques deployed against child violence issues. The trafficking and exploitation of young people has received heightened global attention since the passage of the United Nations protocol on human trafficking in 2000 (UN General Assembly, 2000), which set international standards criminalizing human trafficking for the purposes of labor or sexual exploitation. Feminist scholarship has generally been critical of the language and motivations behind current antitrafficking policies and ideologies. The drafting brought together "strange bedfellows" (Doezema 2005): a peculiar constellation of religious advocates, antiprostitution feminist activists, bureaucrats, and law enforcement. In the nearly two decades since the passage of the protocol, these actors have continued to work in the field they dub "antitrafficking"—simultaneously a social movement, a political moment, and a lucrative career in the human rights field.

The professionalization of antitrafficking has led to an increased presence in non-Western countries of NGOs focusing on saving young female victims of sex trafficking, perpetuating what many have described as a feminization of victimhood and a masculinization of rescue (Andrijasevic 2007; Soderlund 2005) and state support for the neoimperialist rescue of non-Western women, especially those identified as "prostitutes," in the name of humanitarian intervention (Soderlund 2005; Kempadoo and

Doezema 1998; Desyllas 2007). Sex workers have accused antitrafficking advocates of capitalizing on donations meant to combat the exploitation of children to instead target sex workers for moral reasons. The visibility afforded to sex trafficking—as opposed to issues of education, poverty, or environmental justice—has produced its own rescue industry (Agustin 2007) of professionalized humanitarians, journalists, bureaucrats, and corporate representatives seeking to stake their claim on antitrafficking turf.

Drawing from Riles's (2001, 2008) concept of human rights activists forming "antinetworks," I have found that antitrafficking activists are assembling themselves as a "counternetwork," responding to trafficking as knowledge professionals who share jargon, documents, and meeting sites (Thakor 2013). In recent years, publicity campaigns by antitrafficking activist organizations have taken increasingly innovative and high-tech approaches to garner public support for their policy and program agendas. This counternetwork constitutes technosocial solutions to what they conceive as an increasingly complex, digital, infrastructural problem of global human trafficking; the coproduction of solution and problem is an issue I explore in this essay. I argue that Sweetie, in particular, exemplifies the racial exoticization and artifice used, perhaps paradoxically, in antitrafficking campaigns that increasingly blur with formal policing schemes to perform spectacles of rescue.

My analysis of Project Sweetie is part of a larger project on the design and use of new digital techniques to locate child pornography and sexual abuse online, from entrapment schemes to algorithmic filtering of nude images online, to make the argument that new child protection measures mobilize heightened levels of digital surveillance. This essay focuses on Sweetie for its unprecedented use of a particularly racialized and sexualized artificial character—the product of unique collaboration—designed to be a "lure" and entrap potential predators. Based on ethnographic fieldwork[1] with the NGO and creative team behind Project Sweetie, as well as officers from the Dutch national police, I describe how perceptions of digital risk and racial desire critically underpin the production of child protection campaigns. I conducted my fieldwork by procuring an internship at Terre des Hommes at the denouement of Project Sweetie, as the organization was grappling with newfound international publicity for the campaign.

This essay argues that the particular ways in which Sweetie is animated—technologically and socially, through its affective resonance as a publicity campaign—are revelatory of a new development in policing, deception by design. By animating a racialized child avatar as a body that performs on webcam chat rooms and attempts to learn information about the older men that solicit the so-called child, the design team generates a peculiar type of proximity. The Project Sweetie campaign, both the operation and the publicity thereafter, is illustrative of how lures are designed and can operate, and I argue that the design of desirable technology is intimately linked to the design of punishable objects or persons. The team needed to become intimately knowledgeable about the type of potential "offenders" they wished to entrap in what they called a sting operation—who might be intrigued by Sweetie, and who might be likely to divulge personal information in exchange for what they believe to be future sexual promise. Once these names had been acquired and delivered to various law enforcement agencies, the team set up the next stage of the campaign, a curated publicity video, research document, and web presence. This second step, the publicity campaign itself advertising after the entrapment scheme, reinforces and produces Project Sweetie in totality.

Mapping the Problem: "Webcam Child Sex Tourism" in the Philippines

The Netherlands-based Terre des Hommes entered the antitrafficking scene in 2010, following a significant internal restructuring that shifted the organization's focus almost entirely toward addressing the sexual exploitation of children. Its new slogan, *Stopt kinderuitbuiting*, which accompanies all promotional materials, translates literally to "Stop child exploitation." The managerial and organizational changes within Terre des Hommes are representative of a global change to many children-focused organizations in the period following the passage of the Palermo protocols.[2]

Terre des Hommes has long had an interest and stake in Southeast Asia. It maintains regional offices and partners in several major cities in the region, including Jakarta (in general, Dutch NGO presence in the former colony of Java and the Dutch East Indies is quite common), Bangkok, and Manila. Under the new organizational shift, field projects that had been focused on child nutrition, education, and poverty were redirected to amplify cases of child labor and sexual exploitation.

In the Philippines, specifically, child sexual exploitation issues are "very unique" for a confluence of factors, according to Hans Guijt, head of special programs at Terre des Hommes: the prevalence of English as a commonly spoken language, the average lower age of girls and women working in the sex trades, and a sense of familial obligation to earn a contributing income (Guijt, pers. comm.). Additionally, Guijt suggested, "There is almost like an obsession with girls and women for Western men. It's not a universal truth. But the people we work with in the slums, they see that as a way out of their poverty trap. To befriend a Westerner, a rich person, a sugar daddy."

These factors cited by Guijt are explored in detail in a 2012 film, *Lilet Never Happened*, produced by a Dutch filmmaker under consultation with Terre des Hommes's Special Programs division (Groen 2012). Jacco Groen directed this story of a twelve-year-old girl working in a hostess bar in Manila, and the Dutch aid worker woman who attempts to rescue her, based on a real teenager who had come through one of Terre des Hommes's partner programs in Cebu. The film attempts to complicate issues of age and agency, labor versus exploitation, and the politics of rescue. But it also reinforces an ideology of Western male desire and reactionary Asian sexual corruption that I explore in this essay using Anne McClintock's theoretical framing of "the porno-tropics" (1995). The film was positively reviewed and the lead actresses each received several independent awards for their roles; the positive feedback for the film project seemed to have set the stage for Terre des Hommes to become more comfortable with high-publicity endeavors and collaborations with creative arts partners.

By 2011 and 2012, Guijt's team had begun to learn about specific neighborhood communities in the Philippines communally coercing youth to perform in front of webcams. At this point they and many other organizations, as well as the Philippines government, had recognized that certain urban areas had higher proportions of youth (girls and boys both) involved in digital, transactional sex work, especially in Internet cafes. Some were self-motivated, some were employed by Western men or Filipina women, and increasingly, many were coerced by their families. This "cottage industry," as Guijt called it, of multiple families working together to employ or coerce their children into working on chat sites, was increasingly a newer "phenomenon":

> It's an entire neighborhood slum slash neighborhood, where families started to engage in that kind of child sex tourism by putting their own children in front of the webcam and establishing contact with Westerners. At first the mother establishes contact; then she would get feedback like, "Hey, I love you very much, but what about your daughter?" . . . And *much* younger than the kids working from the Internet cafes. These are as young as, say, the client wants them. And if they don't have an eight-year-old boy . . . they knock on the door of the neighbors and say, "Can we borrow your son for twenty minutes?" And they say, "Okay, pay me twenty dollars for my son." So it started off with individual adults, then the entire family became involved, then the entire neighborhood became involved. It became an *industry*—people working from their homes, not from offices or dens or what-have-you. A cottage industry. (Guijt, pers. comm.)

Terre des Hommes launched a two-part intervention onto the "phenomenon," for which they did not yet have a name: research on the psychological effects of digitally mediated sex work on young people; and marketing, through a press trip for Dutch journalists to accompany Terre des Hommes staff through neighborhoods around the cities of Cebu and Cordova in the Philippines. The director of Lemz, a prominent Dutch advertising firm, viewed the subsequent press releases and an article about child sex tourism, and approached Terre des Hommes to discuss the issue. This director, Mark Woerde, saw the potential for a high-profile publicity stunt that would capitalize upon the complex digital elements of the alluring problem of child sexual exploitation, and afford a new marketing opportunity for both the NGO and the ad firm alike.

Animating the Solution: A Virtual Decoy in the Netherlands

Woerde and Guijt decided to label the Philippine phenomenon "webcam child sex tourism" (WCST) to emphasize a particular linkage between Western digital tourism and sexual exploitation in the Philippines. The two men assembled a team from the NGO and the ad firm to begin field research—without video—by pretending to be a young girl messaging people in chat rooms. They operated for three months out of a warehouse in the Netherlands, a location they still will not divulge. "We were extremely paranoid. . . . We didn't know who would come after us once we started getting deep," Zayn, one of the Lemz members explained to me in an interview, articulating the team's apprehensions during the initial preparation for the operation (Zayn, pers. comm.). They feared high-level criminal syndicates would notice and retaliate; for this reason the team of researchers involved was also kept quite small, and many other employees at Terre des Hommes were completely unaware of Project Sweetie until a month before the press release in November 2013.

The chatting team never initiated contact with another person, but always waited to be contacted. They were, however, deliberate about choosing a variety of chat room websites to wait on, from sites that explicitly advertised chats with Filipina women to generalized teenage web chat sites, hoping to encounter a variety of nationalities if possible.

For the bulk of six months of these chats, the iconic, memorably visible Sweetie face was *not* used. This fact is noted in the official Terre des Hommes written report on Project Sweetie, but not featured in any campaign videos or press releases. Marjolijn, a marketing director at Terre des Hommes, explained that technically:

> "Sweetie" did not trace the one thousand men—the face of Sweetie. We only used her a few times, at the end of the project. Because it was very hard to develop her. It's a high-tech Hollywood technique. It took more and more time. This was also kind of frustrating because the research already started, tracing the identities of the men performing webcam child sex tourism. And we also found out we did not *need* Sweetie at that time. Because we were just there in chat rooms. And men were all, they were—if you would just introduce yourself as a ten-year-old from the Philippines, then they already would start very explicit conversations. At that time you did not really *need* a character. (Marjolijn, pers. comm.)

Very early on in the operation, the team realized that the chats with their avatar—varyingly dubbed "Sweetie," "Honey," and "Baby"—could become quite explicit, with people asking for the chatter's age, gender, sexual preferences, whether she would engage in a sexual conversation, and finally, whether she would turn her webcam on for a video show. Zayn confessed he was "freaked out" by how quickly the solicitations came, especially when they posted younger ages for the avatar: "It was stupidly easy to get these guys to give us their email addresses or even names when we offered that Sweetie would do X or Y, or take off her clothes" (Zayn, pers. comm.). Another team member added, "It was unbelievable. . . . It would just be like, *pop-pop-pop*, all the time people asking for chat conversations" (Emma, pers. comm.). The team would pretend that the online connection was poor and offer to send photographs to the chatter's email address. Sometimes the chatters would subsequently discontinue the conversation, but often they would obligingly offer their email addresses.

Capture: Allure, Secrecy, and Trickery

The team decided to go a step further and assume a specific identity for their chat character. Lemz and Terre des Hommes partnered with a graphic design firm specializing in visual effects and facial animation to develop a physical image of a girl for the avatar. Using facial motion capture and a live human actor, the team mapped fifteen unique facial expressions and movements for the avatar.

Dorine, one of the team members who was brought on toward the end of the operation, described the types of "tricks" the team used to elicit information from the men with whom they chatted without giving away Sweetie's true identity:

> Then the guy would reveal his intentions, and they would agree on a price or so. And then she would put on—or, *we* would put on—her webcam. And they had all these tricks to—how to get more information from the guys. They would say, "Oh, the webcam on this messenger service doesn't work! Give me your—do you have a Yahoo address? Maybe we can try it at Yahoo?" And he would give his Yahoo address, and they would have an extra address, or find out his name. So that way they only used the webcam for little bits. So they would never get to a point where she had to put out her clothes. They had all these type of tricks, how to keep the man waiting on the site.
>
> . . . He would put on his webcam, and they want to capture his face, of course—then they had his face. So, yeah, there were a lot of webcam images of just dirty, dirty, naked men . . . and then they would try to get him to put the webcam to his face, so they could capture an image of his

> face. They would just do it quickly—*click!*—but they captured all the images so they could retrieve it. (Dorine, pers. comm.)

The captured faces and names of the men were contrasted against publicly available Facebook or LinkedIn profiles. The correctly matched photographs were printed and posted on a large bulletin board in the warehouse, and eventually printed into a large document packet submitted to the Dutch national police and Europol. The Terre des Hommes marketing director explained: "We made a wall of five hundred of the one thousand men. Unrecognizable, but you could definitely see part of the body. We did that because we knew if there would be media attention, they would want to film. . . . You also want to visualize things, and that picture wall was very impressive" (Marjolijn, pers. comm.). The physicality of the bulletin board and printed packet made the data capture material and tangible.

If people chatting with Sweetie requested further movement or video, the team would again fake a poor connection and request an email address where they might send photos (an offer they would never follow through on). The team emphasized with me that the chats were *never* of a visually sexually explicit nature, despite what the promotional video would later imply. "This was not child pornography. Just an image of a normal girl chatting with guys who revealed their interest in [a] sexual show from Sweetie. But there was no child pornography or anything involved," Emma, a team member at Terre des Hommes, reassured me:

> You just saw her face—you know, the face you see in the media. And she had several moves she could do, like this and type and look up and down. But fully dressed. She was just chatting with the guy. So then the guy would ask her if she would do this and that for so many dollars, and that is when the investigators would try to find out the identity of the guy, so they would try to keep chatting with him as long as possible and get as much information from him as possible. Because he had already admitted that he wanted that and wanted to pay for it. But they would never make it one step further, because it would be illegal. (Emma, pers. comm.)

Sweetie did not "speak" until the campaign video was released—during the sting operation itself, the activists would pretend that the audio connection was broken and conduct all communication by written type. The typing eliminated the problem of finding a live, realistic child's voice to use during the operation. However, the use of the distinctly Filipino-accented child's voice in the publicity campaign suggests another level of artifice in the campaign video's depiction of the sting operation. The campaign's advertisement video suggests perhaps a more "lifelike" simulation than was actually deployed.

The Racial Erotics of Artifice

The virtual decoy of Sweetie was vital to both the legality of the sting operation and the success of the subsequent publicity campaign. The practice of sting operations, or entrapment schemes, is quite commonplace in the US; however, in the Netherlands, and most of Europe, police are forbidden by law from conducting stings using human decoys (adults posing as children). Terre des Hommes, as a civilian group, could not have legally conducted a sting using a human decoy.

The stunning accomplishment of Project Sweetie as a publicity campaign is the design team's self-conscious, stylized acknowledgment of the avatar's artificiality. By conducting Project Sweetie in two parts, as the operation itself and then the subsequent publicity video of the operation, the campaign team is able to specifically frame various elements of Sweetie's artificiality and ethnicity. The Sweetie avatar's face is in fact modeled after actual young people who have come through Terre des Hommes's local service partners in the Philippines. Sweetie is an amalgamation of real children's faces from the photograph database of Fellowship for Organizing Endeavors (FORGE), a community organization based in Cebu City that provides assistance to transient and houseless youth at risk for commercial sexual exploitation or who engage in survival sex. In viewing the Project Sweetie campaign video, the audience does a kind of double seeing: knowing that Sweetie turns out to be a digital model, even vocalizing itself as "not real"—and yet knowing that Sweetie's kind, other similar children, exist in the real world and potentially face actual online harassment and exploitation. In the words of Leonie, a staff researcher I interviewed at the Dutch National Rapporteur's office, "[Sweetie] would appeal to perverted European sex tourists" (Leonie, pers. comm.). This blunt comment was echoed numerous times throughout the course of my research, producing the sentiment that Sweetie was specifically designed as an object of racialized desire.

Artificiality that does not quite reach the point of uncanniness produces affects of cuteness and affinity (Rhee 2013). The work of animation is a collaborative effort; it is also a playful effort. The process of motion capture is the same as that used by many film and video game designers. Additionally, the secretive design process and the concealed "headquarters" warehouse add drama to the designers' sense of their craft. The avatar is animated to perform childish innocence, and femininity, and simultaneously represents adult cunning and artifice. As I suggest, this is precisely the allure of artifice and artificiality. To put it another way, the Project Sweetie operation and campaign have mimetic effects: The viewers of, first, the operation, are lured into the scheme (to collect their personal identifying data) in much the same way that traditional sting operations work. The viewers of the second part of Project Sweetie, the campaign video, can enjoy all of the filmic, shaming qualities of the operation as a whole, and savor the acknowledged artificiality of the character. As Sweetie narrates in the introduction of this essay, viewers of the campaign quite literally follow Sweetie talking and revealing herself—not sexually, but as an artificial creation. I contend that the revelation of Sweetie's artificiality is a process with its own erotic charge. The campaign video forces the viewer to participate in the moment of revelation, a doubly voyeuristic act. The desire to uncover artificiality, measure the degree of realness, also comes from the desire to intensify the gaze to "inspect realism" (Aldred 2011, 4). This hyperbolic, intensified attention to viewing is, Reed and Phillips argue, an eroticizing gesture that also heightens racialized otherness in a character digitally "rendered" such that its race and gender are overdetermined and pronounced (2013, 137). Racial encoding in performance technologies is about producing aesthetically pleasing skin tones, face shapes, characters. By producing a specifically raced avatar, the Project Sweetie design team also erotically encodes and interpellates those who would find it attractive.

Science and technology studies informed by feminist and queer theory take stock of artificiality and reality by exploring where the supposed binary between the two becomes crafted and distorted. Feminist studies of computing, especially, illuminate how computer science theories of robotics and artificial intelligence hinge upon static and monolithic notions of that which is "human" (e.g., Suchman 2008). Such notions come from specific positions of gender, race, and class privilege—yet often remain "mystified" and out-of-frame (Stacey and Suchman 2012) in the final production of the animated character. As Bouldin suggests, "Not only do animators draw upon multiple references for the creation of the animated body, but the body that we, as viewers, experience is also radically hybrid and multiple (particularly the commercially produced animated body)" (2004, 10). Sweetie is the admixture of a white Dutch woman's gestures and expressions used for the purposes of motion capture, and Filipino children's actual faces from the photo repository of a Philippines-based NGO. The avatar enables the campaign viewer to conveniently forget the real children being referenced and enter the filmic enchantment of the avatar video. The queer Black science fiction writer Samuel R. Delany suggests that the practiced sci-fi reader is comfortable with half-truths and half-understandings: stories "should provide the little science-fictional frisson that is the pleasure of the plurality of the sci-fi vision" (2005, 297). The campaign video's moment of Sweetie's skin peeling away to reveal the CGI wireframe grid beneath—a moment of pulling back the curtain, so to speak—only further enhances Delaney's "pleasure of the plurality" (297), and the lure becomes doubly alluring.

Conclusion: Savior Voyeurism and Child Protection in the Digital Porno-Tropics

This particular "project" is unique for its blending of social activism, policing intervention, publicity campaign, and digital cinematography. The campaign was the product of collaboration between an advertising agency, a graphic design company, and a children's charity, all attempting to mimic and even subsume law enforcement operations.

Sweetie is also attractive in a different way, as symbolic of new labor value. The campaign team employs Sweetie to work; the campaign video ascribes further agential capacity to the avatar as herself the entrapper of potential sex offenders. A television news broadcast special from *Kababayan Today*, based in Los Angeles, gave coverage to the Sweetie campaign using the headline "Virtual Pinay 'Sweetie' Catches Sex Predators," evoking not only the avatar's ethnicity but also the sense of Filipino labor

(*Kababayan Today*). In the context of the Philippines, which has the highest number of global migrant workers, the majority of whom are women, this headline combines a sense of nationhood with labor obligation and domestic responsibility. Human responses to digital objects are produced through repeated interactions and orientations toward affinity. Some humans are increasingly comfortable and accustomed to certain kinds of digital labor, rendering such labor invisible. Sweetie as a *product* is a fetish—both an alluring object and a labor commodity fetish.

Once the avatar is set in motion, it performs digital labor—conversational and alluring, but not erotic or sexualized—over enough time that the team who animate it and "speak" (in this case, type) for it can obtain identifying information or data about the avatar's interlocutor. The animated speech is collectively performed by a group and follows a set trajectory with the intention to entrap. The avatar's design, appearance, and dialogue are *oriented* toward outing people. This understanding of racialized and eroticized (digital) labor on behalf of NGOs must also necessarily fall against the backdrop of the labor dynamics of sex work. Sweetie's "work" is both similar and highly contrasted to the survival sex and transactional sex strategies used by the youth she represents. But Sweetie's very purpose is to seek supposed deviants—to be alluring enough to ensnare.

Terre des Hommes's stated intention was not to elicit massive arrests or prosecutions, but simply to emphasize the limits of possibility for police action. In doing so, however, the organization further entrenches the normalization of policing and surveilling action by nongovernmental entities. The men whose names Terre des Hommes collected were allegedly committing an offense in their solicitation of the avatar. In turn, in the campaign video publicizing the sting operation, the viewing audience is also invited to participate in a moment of triumph of rooting out and punishing "predators." This moment is captured in the campaign video by panning over a physical "wall of shame" inside the warehouse (Sweetie 2013).

The triumph of saving "Sweetie" is all the more poignant when undergirded by lasting tropes of the Philippines as a "porno-tropics" (McClintock 1995), a site of both European male sexual desire for young Filipina bodies and European desire for configuring the Philippines as poor, dirty, deviant, and in need of saving. The Philippines is situated by Terre des Hommes as a space of racialized fantasy and sexual deviance, replicated in digital space on webcam chat rooms. Project Sweetie successfully invokes the porno-tropic fantasy such that viewers automatically understand and take for granted its digital version of "webcam sex tourism" and comfortably acquiesce to its argument for the policing and punishment of precrime. With Project Sweetie the potential for punishment was spectacular, a dazzling display of the power of digital surveillance: One thousand people's names were collected and handed off to the police. Since 2013, over fifty people in three countries have been arrested after being identified and located via the Sweetie campaign—including forty-six in Australia alone (Woerde 2014). Most of these people had prior arrest records of possession of child abuse image content. Viewers of the campaign video can feel comforted in the notion that a social problem has been solved technologically, and that potential "predators" have been detained to await criminal prosecution.

That an NGO and advertising firm could conduct a decoy sting operation, and be showered with praise, signals an expansion in the willingness of law enforcement and the antiexploitation network to increasingly blend together. The symbolic protection of children through public displays of punishment helps preserve a sense of stability for the antitrafficking network in the face of increasingly inscrutable and encrypted Internet crimes. The Project Sweetie team strategy of "proactive policing" extends the logic of sex offense punishment by actively encouraging nonpolice to engage in the search for and identification of offenders. By framing digital violence against children as an issue with a technological, carceral solution, the Sweetie campaign encourages other groups to take on community surveillance roles in digital space, patrolling and interpellating online users as criminals and fundamentally transforming the ways in which both "activism" and "policing" are practiced.

Notes

1 The ethnographic data for this essay is drawn from in-depth interviews with staff at Terre des Hommes, the Dutch national police, the office of the Dutch National Rapporteur on Trafficking in Human Beings and Sexual Violence against Children, and Lemz, conducted by the author in Amsterdam, Zoetermeer, and The Hague, Netherlands, between April and September 2014. By request, all names used are pseudonyms excepting the special programs director, Hans Guijt.

2 The three "Palermo protocols" supplement the United Nations Convention against Transnational Organized Crime, a multilateral treaty adopted in 2000. The first, the Protocol to Prevent, Suppress and Punish Trafficking in Persons, Especially Women and Children, is referenced in this essay and is commonly referred to by antitrafficking activists as "the Palermo protocol." The second, the Protocol against the Smuggling of Migrants by Land, Sea and Air (referred to as "the smuggling protocol"), makes important distinctions between the consent involved in people smuggling versus human trafficking. The third protocol, against illicit manufacturing and trafficking in firearms, is referred to as "the firearms protocol."

Works Cited

Agustin, Laura Maria. 2007. *Sex at the Margins: Migration, Labour Markets and the Rescue Industry*. London: Zed Books.

Aldred, Jessica. 2011. "From Synthespian to Avatar: Re-framing the Digital Human in *Final Fantasy* and *The Polar Express*." *Mediascape*. Accessed January 10, 2013. www.tft.ucla.edu/mediascape/Winter2011_Avatar.pdf.

Andrijasevic, Rutvica. 2007. "Beautiful Dead Bodies: Gender, Migration and Representation in Anti-trafficking Campaigns." *Feminist Review* 86 (1): 24–44.

Bouldin, Joanna. 2004. "Cadaver of the Real: Animation, Rotoscoping and the Politics of the Body." *Animation Journal* 12: 7–31.

Delany, Samuel R. 2005. *About Writing: Seven Essays, Four Letters, and Five Interviews*. Middletown: Wesleyan University Press.

Desyllas, Moshoula Capous. 2007. "A Critique of the Global Trafficking Discourse and US Policy." *Journal of Sociology & Social Welfare* 34 (4): 57.

Doezema, Jo. 2005. "Now You See Her, Now You Don't: Sex Workers at the UN Trafficking Protocol Negotiation." *Social & Legal Studies* 14 (1): 61–89.

Dorine [pseud.]. Personal communication, May 16, 2014.

Emma [pseud.]. Personal communication, May 13, 2014.

Groen, Jacco. 2012. *Lilet Never Happened*. London: Spring Film.

Guijt, Hans. Personal communication, May 5, 2014.

Kababayan Today. 2013. "Virtual Pinay 'Sweetie' Catches Sex Predators." November 7. https://wn.com/virtual_pinay_'sweetie'_catches_sex_predators.

Kempadoo, Kamala, and Jo Doezema, eds. 1998. *Global Sex Workers: Rights, Resistance, and Redefinition*. New York: Routledge.

Leonie [pseud.]. Personal communication, September 17, 2014.

Marjolijn [pseud.]. Personal communication, May 19, 2014.

McClintock, Anne. 1995. *Imperial Leather: Race, Gender, and Sexuality in the Colonial Conquest*. New York: Routledge.

Reed, Alison, and Amanda Phillips. 2013. "Additive Race: Colorblind Discourses of Realism in Performance Capture Technologies." *Digital Creativity* 24 (2): 130–144.

Rhee, Jennifer. 2013. "Beyond the Uncanny Valley: Masahiro Mori and Philip K. Dick's *Do Androids Dream of Electric Sheep?*" *Configurations* 21 (3): 301–329.

Riles, Annelise. 2001. *The Network Inside Out*. Ann Arbor: University of Michigan Press.

———. 2008. "The Anti-network: Private Global Governance, Legal Knowledge, and the Legitimacy of the State." *American Journal of Comparative Law* 56 (3): 605–630.

Soderlund, Gretchen. 2005. "Running from the Rescuers: New U.S. Crusades against Sex Trafficking and the Rhetoric of Abolition." *NWSA Journal* 17 (3): 64–87.

Stacey, Jackie, and Lucy Suchman. 2012. "Animation and Automation—The Liveliness and Labours of Bodies and Machines." *Body & Society* 18 (1): 1–46.

Suchman, Lucy. 2008. "Feminist STS and the Sciences of the Artificial." In *The Handbook of Science and Technology Studies*, edited by Edward J. Hackett, Olga Amsterdamska, Michael Lynch, and Judy Wajcman, 139–164. Cambridge, MA: MIT Press.

Sweetie. 2013. "Stop Webcam Child Sex Tourism!" YouTube video, 7:42. https://www.youtube.com/watch?v=aGmKmVvCzkw.

Thakor, Mitali. 2013. "Networked Trafficking: Reflections on Technology and the Anti-trafficking Movement." *Dialectical Anthropology* 37 (2): 277–290.

UN General Assembly. 2000. Protocol to Prevent, Suppress and Punish Trafficking in Persons, Especially Women and Children, Supplementing the United Nations Convention against Transnational Organized Crime. Palermo and New York: UN General Assembly.

Woerde, Mark. 2014. "UN congratulates LEMZ on the global impact of 'Sweetie' campaign." Lemz. Last modified February 18, 2014. http://lemz.net/wp-content/uploads/2014/06/140526-Sweetie-announcement.pdf.

Zayn [pseud.]. Personal communication, May 7, 2014.

Gone, Missing

Queering and Racializing Absence in Trans & Intersex Archives

HILARY MALATINO

WHO APPEARS IN THE MEDICAL ARCHIVES THAT DOCUMENT THE treatment of intersex and trans subjects? Who is missing? What can we learn from these absences?

This chapter takes on these questions, examining the phenomena of patient disappearance and the broader archival absence of queer folks and folks of color in the archives of US-based sexologist John Money, who served as one of the primary architects of trans and intersex medical pathologization (see Rubin, "'An Unnamed Blank that Craved a Name,'" this volume). I read these absences and disappearances as a method of resistance to the imposition of what María Lugones has termed the modern/colonial gender system (2007), and explore the implications of the fact that folks whose desires for transformation run counter to hegemonic, white, bourgeois understandings of masculinity and femininity were systematically prevented from accessing technologies of transition, deemed unacceptable candidates or noncompliant patients. By "technologies of transition," I mean the ensemble of medical practices utilized in the process of transition, both hormonal and surgical. This exclusionary, highly regulated system of medical gatekeeping has prompted two linked phenomena: a legacy of trans folks tailoring experiential narratives—lying—to fit the heterosexist, highly binarized logic utilized by classic models of transsexuality, as well as a history of intersex folks critiquing the pathologization of intersex traits and refusing to be interpellated by the medical establishment as disordered, in need of surgical and hormonal gender normalization.

Racialized, classed, and queer absences are central to understanding how access to technologies of transition have become intensively compromised for poor folks, trans folks of color, and gender-nonconforming, nonheterosexual folks while they have, simultaneously, been coercively imposed on intersex folks in the interest of normalizing our divergent forms of sexed embodiment. This selective utilization of technologies of transition helps explain why it is we currently lack holistic approaches to trans and intersex health that move beyond surgical and hormonal techniques of gender normalization and focus more heavily on the structural violence that too often compromises the life chances of trans and intersex subjects.

Repairing Gender: Tactics of Medical Normalization

John Money opened the first clinic in the United States that specialized in intersex and trans conditions—the Johns Hopkins Gender Identity Clinic. This clinic began performing gender reassignment surgery on trans and intersex folk in 1966, and enabled Money to become one of the first sexologists to engage in the substantive study of trans and intersex conditions, treatments, and long-term outcomes. In 1968, he published *Sex Errors of the Body and Related Syndromes*; in 1969, he coauthored *Transsexualism and Sex Reassignment*. These books laid the theoretical and practical groundwork for contemporary forms of intersex and trans diagnosis. They were rapidly translated; Money quickly became an international authority on intersex and trans counseling, surgery, and continued care. He also coined the term "gender role"; our contemporary conceptual habit of separating sex from gender is rooted, in part, in his work.

I approached his archives skeptically. As an intersex person, I am deeply critical of his treatment protocol, which advocates binary gender assignment and genital reconstructive surgery. These surgeries were (and are) performed on infants and toddlers nonconsensually, because Money believed gender identity was solidified before the age of three and that, after that, it would be too psychologically destabilizing for intersex folk to grow up with atypical genitalia. He argued that "gender role is so well established in most children by the age of two and one-half years that it is then too late to make a change of sex with impunity" (Money, Hampson, and Hampson 1955b, 290).

For Money, a "normal" vagina was one capable of intromitting an average-sized penis. This was his gold standard for a well-realized genital reconstruction, and it always, in the case of intersex patients, came coupled with the surgical removal of the clitoris (considered "abnormally large," too phallic for someone being reared as female). He argued that this was an A-OK practice that didn't compromise sexual sensitivity in the least:

> There has been no evidence of a deleterious effect of clitoridectomy. None of the women experienced in genital practices reported a loss of orgasm after clitoridectomy. All of the patients were unanimous in expressing intense satisfaction at having a feminine genital morphology after the operation. . . . There is considerable evidence that an amputated clitoris is erotically sensitive. (295)

It is apparent that, for Money, gender-typical aesthetics trumps erotic functionality. His musings, here, also raise an important question: How can something missing—the amputated clitoris—be "erotically sensitive"? While I certainly don't buy Money's assertion that a missing clit is erotically sensitive, I do want to think about the effects of other forms of absence in relation to coercive forms of medicalized transition: patients who go missing, who refuse to show up for medical appointments, as well as those beings who can't get their foot in the door of the clinic because they're too poor, too queer, too gender-nonnormative.

Money argued the importance of referring to vaginas, reductively, as "baby tunnels" (Money, Hampson, and Hampson 1955b, 295) when discussing anatomy with intersex children. His logic was, if you avoid mentioning the clitoris, intersex children won't realize it exists, and thus won't be upset that they had theirs removed! He also advocated years of invasive continued care, including regular manual dilation of the reconstructed vagina by the child's caretakers (until they were old enough to do so themselves) and a lifelong hormonal regimen meant to "normalize" secondary sex characteristics. He encouraged doctors and family members to lie to the child about their intersex condition. Parents were encouraged to go to great lengths to obscure the realities of invasive forms of gendered normalization, and to do so in the name of protecting intersex folks from the supposedly devastating truth of their own nonconventionally sexed embodiment. As if we wouldn't figure it out!

The literature within the growing field of intersex studies, not surprisingly, has unequivocally critiqued the medical protocol developed by John Money. Anne Fausto-Sterling, in her landmark volume *Sexing the Body,* reviews the modern medical management of intersexuality and concludes, with no uncertainty, "stop infant genital surgery" (2000, 79). Suzanne Kessler, in *Lessons from the Intersexed,* asserts that his method of treatment implements a medical fix—surgical and hormonal gender normalization—for what is actually a social dilemma: a body that doesn't fit neatly within the parameters of sexual dimorphism, our prevailing narrative of biological sex differentiation (1998). Alice Dreger, in her history of the medicalization of intersex conditions, examines the contemporary autobiographical writing of adult intersexuals and concludes that, "despite the effort to make intersexed children look and feel 'normal,' the way intersexuality is treated by doctors in the United States today inadvertently contributes to many intersexuals' feeling of difference and defectiveness" (2000, 190). More recently, Elizabeth Reis has written that "ever since the early nineteenth century, when doctors began to professionalize and publish their cases in medical journals, we can trace not only their cruelly judgmental descriptors of these conditions and people but also the damaging therapeutic treatment they have dispensed" (2009, 157). Rounding out this chorus of academics against the medical pathologization of intersex conditions is Georgiann Davis, a sociologist as well as someone with an intersex trait, who asserts that gendered expectations are what "force intersex people, who do not fit neatly into the gender structure, to undergo medically unnecessary and irreversible surgeries that . . . may be intended to help but are often quite harmful" (2015, 157).

Relatedly, within trans studies, there is a growing number of scholars insisting on grappling with the harm caused by forms of medical gatekeeping that green-light transition only for folks who provide straightforward life narratives that rely heavily on gender stereotypes to claim they are trapped in the wrong body, have a high likelihood of effectively passing as cisgender post-transition, and can attest to their lack of queerness—that is, promise exclusive heterosexual behavior post-transition and demonstrate both chastity and sexual disgust with their bodies pretransition. Historian Joanne Meyerowitz summarizes, in brief, the foundational premises of a more recent generation of trans scholars:

> As a group they tended to start with [the premise] that sex, gender, and sexuality represent analytically distinct categories, that the sex of the body does not determine either gender or sexual identity, that doctors can alter characteristics of bodily sex. Some disputed binary definitions of biological sex . . . many combined the feminists' critique of the constraints of rigid gender dichotomies and the gay liberationists' goal of freedom of expression, and rendered healthy the variations that doctors had routinely cast as illness and disorder. (2004, 284)

Meyerowitz's intellectual touchstones, here, are scholars, activists, and performers like Susan Stryker, Sandy Stone, Kate Bornstein, Henry Rubin, Jason Cromwell, and Riki Anne Wilchins—names now associated with the emergence of both trans studies and contemporary transgender activism. Critiques of binary gender, rejection of the pathologization of trans embodiment as a form of deviance or disorder, and deconstructions of the disciplinary regulation of access to technologies of transition are integral to the field, taken up across the landscape of contemporary trans scholarship.

Legal theorist Dean Spade, for instance, offers an intimate account of his failure to provide the correct gender narrative to psychiatric professionals as he was seeking access to technologies of transition:

> From what I've gathered in my various counseling sessions, in order to be deemed real I need to want to pass as male all the time, and not feel ambivalent about this. I need to be willing to make the commitment to "full-time" maleness, or they can't be sure that I won't regret my surgery. The fact that I don't want to change my first name, that I haven't sought out the use of the pronoun "he," that I don't think that "lesbian" is the wrong word for me, or, worse yet, that I recognize that the use of any word for myself—lesbian, transperson, transgender butch, boy, mister, FTM fag, butch—has always been/will always be strategic, is my undoing in their eyes. They are waiting for a better justification of my desire for surgery, something less intellectual, more real. (2003, 21–22)

Spade highlights the difficulty medical professionals have in cognizing forms of gender variance that don't subscribe to classic narratives of transsexuality, and articulates the realization of his own naiveté in believing that he could actually communicate the complexities of his desires to medical professionals invested in these narratives. He becomes involved with a trans support group, and realizes that this space—not the doctor's office—is open to honest, complex, and mutable truths about queer forms of trans embodiment: "I have these great, sad conversations with these people who know all about what it means to lie and cheat their way through the medical roadblocks to get the opportunity to occupy their bodies in the way they want" (2003, 23).

While Spade excoriates the medical establishment as a regulatory system deeply invested in stereotypical binary gender, he also complicates transnormative narratives of transition that are invested in the reification of hegemonic medical constructions of transition as a linear, teleological path (from male to female, or female to male). By *transnormative*, I mean subjects who, save for their status as trans, are otherwise highly assimilable—gender-normative, heterosexual, middle-class, well-educated, white. It is transnormative subjects who populate the medical archives of transsexuality most heavily, and it is transnormative subjects who have the least mitigated access to medical technologies of gender transition—hormones, surgery, and continued care. Conversely, it is non-transnormative subjects who are systematically exposed to institutional and interpersonal violence, up to and including death—by homicide and suicide, yes, but also by lack of access to quality, affordable, trans-competent health care.

I understand the utilization of technologies of transition by both intersex and trans folks as forms of gender transition. While the medical rhetoric surrounding the surgical and hormonal normalization of intersex folks frames these procedures as correcting or repairing an "unfinished" (that is, not unequivocally male- or female-typical) form of sexed embodiment, these experiences are lived as major reconfigurations of one's gendered reality. With this in mind, I apply the concept of transnormativity to both intersex and trans experiences, using it as a shorthand means of indexing narratives that embrace gender stereotypes, understand technologies of transition as a means of deliverance into the promised land of gender normativity, and utilize heterosexuality as a means of shoring up and verifying gender normativity. We find transnormative narratives of gendered becoming in both contemporary accounts of transition as well as in the archives of medical sexology, issued by intersex and trans subjects alike—although, as Spade points out so eloquently above, often under conditions of coercion, as a means of playing into and verifying the medical establishment's investment in gender and sexual normativity.

Popular accounts of gendered becoming rely heavily on what I understand as *transnormative structures of feeling.* The most popular, palatable narratives of transition confirm gender-stereotypical truisms regarding the biology and psychology of sex difference. The narratives offered up most consistently within John Money's case studies do the same sort of confirmation work, shoring up his idea that gender is a matter of indelible psychological imprinting at a young age. His preferred examples are those that testify to indelibly male-typical or female-typical gender roles, and he relies on narrations of childhood memory that conform to transnormative structures of feeling. For instance, one of the few patients cited at length in his case studies—a male-identified intersex person with hypospadias who had been reared as male—recounts,

> "I remember myself squirting the hose. I think I squirted my father in the process. And it was lots of fun." This memory, dating from the age of two, had been reinforced from a photograph taken on the occasion; "and yet, when I look at it, it kind of brings me back, you know, that wonderful feeling of power you have when you're watering something! Well, it kind of brings back something of being a master in your own domain as you squirt this blasted hose around." (Money, Hampson, and Hampson 1955a, 313)

Not only does Money's paradigmatic patient enjoy the rush of power that comes with squirting the hose, he also explains that he is heavily preoccupied with sex, but only the strictly heterosexual variety, while also manifesting some concern about being too sex-obsessed: "I would think about all kinds of sex, what kinds there were, and then I would wonder if I was safe or not. Whether I would find myself liking it too much or something. Yet, if I ever started making any image of homosexuality, I could never get myself into it" (315).

This narrative shores up the dominant expectations regarding male sexuality of that era—a man was a person who had to, for the social good, rein in sexual drive and make sure his preferred objects were exclusively of the opposing gender. It is no surprise, then, that Money—in his assumed role as arbiter of the truth of gender identity—declares "all in all, beyond every possible doubt, this person was psychologically a man. He was fortified with a diplomatic arrogance which adjusted to the human demands of the occasion, yet enabled him to choose and select his standards rather than run with the herd" (318). Diplomatic arrogance! Pronounced individualism! A natural-born leader, this patient, possessed of all the hallmarks of rugged American machismo. We need no further proof of the immutability of psychological gender, even in cases of ambiguously sexed individuals. In other words, all intersex people have an "indelibly imprinted" psychological gender; it is the job of the medical sexologist to discern what that is and green-light technologies of transition accordingly. His function is, then, only one of enhancing the potential of intersex and trans persons to live unremarkable, gender-typical, and sexually normative lives.

One of the great risks of late gender assignation, for Money, is the possibility of queer sexuality. In a 1965 article, he writes,

> after a child has entered school, a sex reassignment is extremely perilous psychosexually and is liable to produce a person who lives socially in the reassigned sex but falls in love as a member of the other originally assigned sex—and thus has all the outward appearances of being homosexual. These late sex reassignments also may issue in nonspecific, moderate to severe psychopathology of the personality. (187)

To put it bluntly: Not only does Money think late gender reassignment might make you queer, it may also produce mental disorders. This rigorous effort to guard against any association with queer sexuality is not only present in intersex cases, but—as mentioned above—in trans cases as well. The aim of intersex and trans medical treatment was—and in many cases, remains—the production of gender-normative heterosexuals.

Trans Necropolitics and Archival Absence

What if technologies of transition sometimes make us feel like shit, but we utilize them because of the more intense social cost of not being passable as cis? What if we reject hormonal therapy? What if we have reservations about submitting to technologies of gender normalization? What if we can't afford or can't geographically access technologies of transition?

What if transition, for intersex and trans folk alike, is not always a triumphal narrative, but instead a sort of necropolitical calculus wherein negative effects are weighed one against the other?

Trans necropolitics, as theorized by C. Riley Snorton and Jin Haritaworn (2013), refers to the exposure to violence, debility, and death that shapes nontransnormative lives. Trans necropolitics is a useful concept in thinking through how intensely stratified access to medical care is for trans and intersex subjects, how the rise in trans visibility comes coupled with an intensification of violence toward nontransnormative subjects, and how the livability of transnormative lives is interwoven with institutional mechanisms that expose less privileged trans and intersex subjects to systemic violence and disenfranchisement.

Can we understand decision-making that takes place within this milieu an instance not of willed, autonomous self-making, but instead as consent compromised by conditions of coercion? Why have we not paid more attention to the historical incidences of folks who experience trauma in their interface with medical professionals, rather than validation? Those who reject the notion that medical specialists are also, somehow, saviors? What about the experience of affect aliens who, as Sara Ahmed writes, "do not experience pleasure from proximity to objects that are attributed as being good" (2010, 41)?

I began thinking about this while working on a project documenting my own experience growing up intersex. I was diagnosed late—around the age of sixteen—and promptly put on Premarin, a conjugated estrogen that was, at the time, commonly used to treat intersex conditions, as well as for hormone therapy in both cis and trans women. This was meant to normalize my intersexed body along female-typical lines, resulting primarily in breast growth and fat redistribution. After months of severe depression, including suicidal ideation, I stopped taking it, and haven't been on hormones since. It's difficult, existentially speaking, to tease apart the side effects of that particular drug from the general trauma of grappling with an intersex diagnosis, but I do know that my decision to cease hormonal treatment was directly linked to a substantial decrease in the intensity of depressive symptoms I experienced. I also understood my refusal of hormone therapy as a refusal, more broadly, of medical tactics of gender normalization aimed at intersex youth and adults; a refusal of the notion that my corporeal queerness needed to be fixed or remediated.

When, some years later, I visited the archives of the medical sexologists who produced the treatment protocols I'd been subjected to—and run away from—I found my experience mirrored, although it was obfuscated by the curatorial impulses of these medical professionals, veiled by their desire to protect and render watertight their theories. I discovered anger on the part of intersex patients over the trauma they experienced within the medical establishment. Patients repeatedly refuse to return for further examination, finding their own gendered and sexual inclinations at odds with Money's recommendations for treatment. Several patients, when slated to have their genitalia photographed by Money's assistants for medical publications, simply don't show up. One patient—an androgen-insensitive person reared as male whom Money, in concert with this boy's parents, insisted on reassigning as female—went so far as to call Money on the coercion evident in his treatment methods:

> I think you're a rotten guy. I told my father that you were trying to make me do what he wants. And I think the same thing of you. . . . You're trying to make me say what you want me to say. And I don't want to say that. . . . I told you what I want. You said we won't mention nothing about the other sex that I don't want to be. . . . What you're saying is to imagine that I'm the other sex, that's what you're saying, and I don't like imagining that way. (Money 1991, 45)

What is shocking is that Money doesn't interpret this anger as directed at him, as stemming from the patient's profound disagreement with his dogged insistence on gender reassignment. Rather, he believes it is the product of a wrongly assigned gender, believes this child is upset because he was reared as male despite being intersex, and thus not possessing what, for Money, was the ultimate arbiter of manhood—a "normal-looking" penis. Money concludes the case history noting the patient's eventual suicide, which he argues would not have happened had this patient heeded his advice and accepted gender reassignment as female. This misreading of patient affect is consistent across Money's case studies, and prompts my concern that, in trans and intersex narratives alike, the elements of coercion involved in medical procedures of gender normalization have been significantly downplayed.

Historically speaking, folks whose desires for transformation run counter to hegemonic, white, bourgeois understandings of masculinity and femininity were systematically prevented from accessing technologies of transition. The forms of gender normativity utilized by the medical establishment were—and remain—undergirded by race, insofar as what was understood as a normative gender ideal was implicitly white, shaped by the typologies of masculinity and femininity that apply to what decolonial feminist philosopher María Lugones has called the "light" side of the colonial/modern gender system.

Lugones reasons that white bourgeois ideals of gender embodiment have been shaped by a deeply dimorphic understanding of gender complementarity that emphasizes white female sexual submissivity, domesticity, and minimized agency and access to the public sphere, and white male providence, epistemic and political authority, virility, and naturalized dominance. This "light side" of the colonial/modern gender system stands in contrast to a "dark side," constituted by the ways in which the sexualities, embodiments, and kinship forms of colonized peoples were constructed within the colonial imaginary. As Lugones writes in "Toward a Decolonial Feminism," "the hierarchical [gendered] dichotomy [that characterizes the 'light' side] also became a normative tool to damn the colonized. The behaviors of the colonized and their personalities/souls were judged as bestial and thus non-gendered, promiscuous, grotesquely sexual, and sinful" (2010, 745). In short, the gendered norms and mores that have determined the *telos* of biomedical logics of gender transition are also those that have been utilized to frame the kinship forms, sexualities, and embodied intimacies of peoples with legacies of colonization as aberrant and in need of rehabilitation and assimilation.

Emily Skidmore, in her media analysis of mid-twentieth-century representations of transwomen, argues that those women with the most "proximity to bourgeois white womanhood" were represented most frequently, and their stories "came to define the boundaries of transsexual identity" (2011, 271). Moreover, access to technologies of transition was, and remains, doubly compromised for trans folks of color; as Delisa Newton attests, in a 1966 issue of *Sepia* cited by Skidmore,

> Because I am a Negro it took me twice as long to get my sex change operation as it would have a white person. Because I am a Negro many

> doctors showed me little sympathy and understanding. "You people are too emotional for such an ordeal," one doctor told me. But finding medical attention wasn't the only problem complicated by the color of my skin. Even with my college and nursing education, I couldn't get a good, steady job to raise money for the operation. (292)

My own work in the Kinsey archives verifies this phenomenon of compromised access, which manifests most often as archival absence. Trans and intersex folks of color are conspicuously missing from the medical archives of sexology; moreover, many folks—white folks and folks of color—appear briefly in medical records, only to never return, in effect going AWOL from the medical protocols of transition and gender normalization. Despite this, never in the work or correspondence of either of these massively influential sexologists I've researched was there any reflection on the partiality of knowledge manifest in such a racially homogenous, Westocentric archive. In the rare moments that folks of color appear in these archives, they are framed, in accordance with the logic of the "dark side" of the colonial/modern gender system, as deviant, sexually perverse, and culturally both aberrant and anachronistic.

For example, take the image of an indigenous American—a member of the Diné people—that I found in a box of photographs marked "intersex" in the Kinsey archives. This person may or may not have had an intersex condition, but was more possibly *nadleeh*—a Diné conception of embodiment that is not accurately translatable into Western gendered logics, although it is often referred to as a type of third gender. This was the only photograph of a nonwhite subject in that box, as well as the only photograph that was not formally composed and set indoors, in a photo studio or medical clinic. The text beneath the photo reads:

> A Navajo Indian. Age 27. Height 5'7", weight 150, length 2.0", diameter 0.3". No hair on body and no sign of testes. Scrotum contained only a soft mass of indistinguishable tissue. Erection and orgasm possible but orgasm slight with emission of a few drops of what appeared to be semen. Intelligent and normal in other ways. He had attended Indian Boarding School. Was rejected by army draft board because of his sex organs. He tries coitus and enjoys it. Gets most satisfaction with little girls, but prefers adult women. They ridicule him because of the size

> of his organs. He feels his condition deeply, and begged to be told how he could "make it grow, so he could get married and have babies." His concern was over the size of his penis, not seeming to attach much importance to the lack of testes. He is probably one of those individuals who some tribes develop for pederasty through non-instrumental castration while small boys, although he denied it. If he is, he apparently has rebelled and desires to be normal. They are usually very effeminate in appearance and actions, but he was not. He is experienced in fellatio and pederast, the anal muscles being quite relaxed. Adult male organs attract him very much and he delights in handling and gazing at them. He is particularly fascinated by semen, which, however, is not unusual in Indians. He was reluctant to pose which, combined with lack of seclusion, prevented more and better pictures. (Kinsey Institute Archives, n.d.)

This man is framed as living proof of the sexual and gendered deviance of the Diné people; he is presented as both irrefutably perverse in relation to Western gendered and sexual norms, engaging in nonheterosexual, age-inappropriate sexual activities, but also as victimized by the ostensibly strange sexual customs of the Diné and desperate for the forms of gendered normalization Western medicine can provide. His desire for gendered and sexual "normalcy" is implicitly linked to his time spent in the viciously assimilatory Indian boarding school system. White, Westocentric gendered and sexual normalcy is aspirational for this person; the medical specialist is simultaneously the gatekeeper and the benevolent colonial patriarch, able to make these dreams come true.

Racialized, classed, and queer absences and misrepresentations of this sort are central to understanding how access to technologies of transition have become compromised for poor folks, folks of color, and gender-nonconforming and queer folks. Popular understandings of trans and intersex identity are linked indissolubly to medicalized transition. Access to medicalized technologies of transition is too often understood as the sine qua non of trans and intersex livability and health. We are in dire need of holistic approaches to health that move beyond surgical and hormonal techniques of gender normalization and focus, instead, on remediating the quotidian and structural violence that so often compromises the life chances of trans, intersex, and gender-nonconforming subjects. It is imperative to interrogate this exclusionary legacy of medical treatment as the transnational market for medicalized transition grows while the communal, nonprofit networks of support, advocacy, and assistance that are able to address the exigent needs of trans and gender-nonconforming subjects remain relatively stagnant.

Works Cited

Ahmed, Sara. 2010. *The Promise of Happiness*. Durham: Duke University Press.

Davis, Georgiann. 2015. *Contesting Intersex: The Dubious Diagnosis*. New York: NYU Press.

Dreger, Alice Domurat. 2000. *Hermaphrodites and the Medical Invention of Sex*. Cambridge, MA: Harvard University Press.

Fausto-Sterling, Anne. 2000. *Sexing the Body: Gender Politics and the Construction of Sexuality*. 1st ed. New York: Basic Books.

Kessler, Suzanne J. 1998. *Lessons from the Intersexed*. New Brunswick: Rutgers University Press.

Kinsey Institute Archives. n.d. Untitled photograph of member of Diné tribe. Medical Box Number 56634, Folder "Intersex." Kinsey Institute Archives. Indiana University Libraries, Bloomington.

Lugones, María. 2007. "Heterosexualism and the Colonial/Modern Gender System." *Hypatia* 22 (1): 186–219.

———. 2010. "Toward a Decolonial Feminism." *Hypatia* 25 (4): 742–759.

Meyerowitz, Joanne J. 2004. *How Sex Changed: A History of Transsexuality in the United States*. 1st paperback ed. Cambridge, MA: Harvard University Press.

Money, John. 1965. "Psychology of Intersexes." *Urologia internationalis* 19 (1–3): 185–189.

———. 1968. *Sex Errors of the Body and Related Syndromes: A Guide to Counseling Children, Adolescents, and Their Families*. 2nd ed. Baltimore: John Hopkins University Press.

———. 1991. *Biographies of Gender and Hermaphroditism in Paired Comparisons: Clinical Supplement to the Handbook of Sexology*. Amsterdam: Elsevier Science Ltd.

Money, John, and Richard Green. 1969. *Transsexualism and Sex Reassignment*. Baltimore: Johns Hopkins University Press.

Money, John, Joan G. Hampson, and John L. Hampson. 1955a. "An Examination of Some Basic Sexual Concepts: The Evidence of Human Hermaphroditism." *Bulletin of the Johns Hopkins Hospital* 97 (4): 301–319.

———. 1955b. "Hermaphroditism: Recommendations Concerning Assignment of Sex, Change of Sex and Psychologic Management." *Bulletin of the Johns Hopkins Hospital* 97 (4): 284–300.

Reis, Elizabeth. 2009. *Bodies in Doubt: An American History of Intersex*. Baltimore: Johns Hopkins University Press.

Skidmore, Emily. 2011. “Constructing the ‘Good Transsexual’: Christine Jorgensen, Whiteness, and Heteronormativity in the Mid-Twentieth-Century Press.” *Feminist Studies* 37 (2): 270–300.
Snorton, C. Riley, and Jin Haritaworn. 2013. “Trans Necropolitics.” In *The Transgender Studies Reader 2*, edited by Susan Stryker and Aren Z. Aizura, 66–76. New York: Routledge.
Spade, Dean. 2003. “Resisting Medicine, Re/modeling Gender.” *Berkeley Women's Law Journal* 18 (1): 15–39.

PART THREE

Disruptive Practices

THE FIVE ESSAYS IN THIS SECTION ANALYZE OR IMAGINE DISRUPTIVE practices. In brief, Michelle Murphy analyzes the efforts of feminist women's health activists to disrupt normative understandings of cervical health; Rachel Lee analyzes comedian Margaret Cho's use of peristaltic feminism to disrupt Western imperialism; Kane Race draws on existing practices of gay men to imagine a disruptive HIV/AIDS harm reduction strategy that focuses on pleasure and embodiment; Amber Musser uses the writings of Gary Fisher to imagine a way of thinking about sexual consent that disrupts legal demands for "able-mindedness"; and Isabelle Dussauge engages in the utopic imagining of a future in which a disruptive past has created possibilities for a "queer neuroscience."

We want to be clear that all of the essays in *Queer Feminist Science Studies: A Reader* are "disruptive projects" in different ways. For example, in critiquing science and medicine, many of the pieces in this volume disrupt hegemonic ways of thinking about the "stuff of the world." What differentiates the essays in Part Three, then, is that not only do they engage in disruptive analytic processes, but they take disruption itself as an object to analyze and/or imagine. Unifying these essays is the Foucauldian and queer feminist science studies recognition that disruptive practices can successfully challenge and imagine alternatives to hierarchical systems of oppression and domination, but that disruptive and alternative practice are never outside of or innocent of the systems in which they operate. In Rachel Lee's words, there is "no subsumption and no transcendence but only peristaltic waves that push (or connect us to) the world through both ends."

Michelle Murphy's "Immodest Witnessing, Affective Economies, and Objectivity" comes from her 2012 book *Seizing the Means of Reproduction*. The book examines health practices developed by radical feminists in the US in the 1970s and 1980s (sometimes called "the feminist women's health movement"), identifying both radical effects and entanglements with

American empire, population control, and neoliberalism. The piece included here examines the iconic practice of vaginal self-exam, arguing that this practice refashioned objectivity by materializing the researcher as particular, embodied, and affectively entangled with her object of study. At the same time, the practice was implicated in broader capitalist trends toward the use of "affect" in marketing and the development of profitable reproductive biotechnologies. Murphy's piece serves a dual role in this reader: simultaneously identifying members of the feminist women's health movement as early queer feminist scientists and, herself, demonstrating a practice of queer feminist science studies in her attention to regimes of normalization, emancipatory politics, and the ways in which emancipatory projects are always also implicated with broader systems of power.

Rachel Lee's "Pussy Ballistics and the Queer Appeal of Peristalsis, or Belly Dancing with Margaret Cho," first published in *GLQ* in 2014, examines the biopolitical and geopolitical interventions into racialized gender and sexuality that Korean American stand-up comedian Margaret Cho's comedy literally performs at the site of the gut, first registering as belly-shaking laughter before becoming processed cognitively and intellectually. Lee's concept of pussy ballistics refers to the vaginal explosions that at once punctuate, interrupt, and call into question US imperialism, militarism, and biomedical science through Cho's comedy. As a meditation on how biology offers alternative models of embodiment and potentialities for disruption that might be useful for queer, feminist, and antiracist critique, Lee powerfully demonstrates the relevance of "gut feminism" (Wilson 2015) to efforts to think natureculturally and developmentally in ways that refuse teleological narratives of closure and redemption. Instead, Lee calls on feminist and queer students and scholars to inhabit the kinesthetics of what it feels like to laugh at and in the face of the profound (and profoundly absurd) inequalities, corporeal and discursive violences, and forgettings of the history of the present.

Kane Race's "Embodiments of Safety" comes from his 2009 book, *Pleasure Consuming Medicine: The Queer Politics of Drugs*. The book as a whole examines discourses surrounding drug use (e.g., "treatment vs. recreation" and "licit vs. illicit") and what function these discourses serve for the neoliberal state. In the essay included here, Race examines how discourses about gay men and illicit/recreational drug use position

Australian gay men as hedonists irresponsibly courting HIV/AIDS infection. As an alternative, Race analyzes the disruptive, embodied, and pleasure-attentive harm reduction strategies that have been developed by some gay men and public health practitioners. The essay is a powerful example of scholarship that simultaneously takes into account broader operations of neoliberal power and local embodied practices, while both critiquing moralistic public health efforts and offering an alternative, disruptive model of public health based on an appreciation of embodiment and pleasure.

Amber Musser's "Consent, Capacity, and the Non-Narrative" is an original contribution. Musser analyzes two cases in which sexual consent was called into question—one in which a man was charged with rape for having sex with his wife who had advanced dementia, and one in which a male college student was expelled for sexual misconduct after he and a female student had sex while they were both intoxicated. The cases highlight the limitations of a notion of "consent" that can never be achieved by certain subjects. To think consent in a way that disrupts the legal demand for able-mindedness, Musser turns to the journals of gay African American writer Gary Fisher, who died of AIDS at the age of thirty-two. Musser's piece does important critical work in deconstructing consent in addition to offering a disruptive and imaginative way to rethink sexuality athwart notions of pathology, coherent linear selfhood, and capacity. Her piece thus draws on and contributes to a rich history of feminist, queer, intersectional, and disability studies scholarship on gender, pathology, consent, and sexuality.

The final essay in this section is Isabelle Dussauge's "Brains, Sex, and Queers 2090: An Ideal Experiment," in which Dussauge projects herself into the future (2090) and narrates the "history" of 2012 to 2090, during which time the once-disruptive *neuroQueer* movement successfully revolutionizes society, producing a queer utopia. Dussauge's essay invites readers to imagine a world in which queer science has become so much a part of the fabric of science that its origins are of interest to only a few historians. The tone of the essay is playful, including helpful notes about the oddity of scientific beliefs of the early twenty-first century, a list of fake references, and allusions to fictional historical events. Dussauge focuses on the design of the first truly queer neuroscience experiment: a "data-crowded," multigender examination of desire entitled "Power, Brains, and Sexuality," published in the (fictional) journal *Mature Neurosciences* in

2019. Ultimately, her piece denaturalizes current paradigms while also serving as an example of a rhetorical paradigm in which a call for radical new sciences includes thinking, first, about where we want to be, and then how to get there from here.

Discussion Questions

1. Are there any commonalities between the disruptive practices discussed in these five essays (and, if so, what are they)? What are the unique elements of these disruptive practices?
2. The essays in Part Three suggest that disruptive practices are always, in some ways, complicit with the systems in which they operate. What does this mean, ethically and politically, for those attempting to develop alternatives to hegemonic practices of science and medicine?

Works Cited

Wilson, Elizabeth A. 2015. *Gut Feminism*. Durham: Duke University Press.

Immodest Witnessing, Affective Economies, and Objectivity

MICHELLE MURPHY

SIT IN A CIRCLE. ASSEMBLE A KIT COMPOSED OF MIRROR, LIGHT, and plastic speculum for each participant. Lubricate your speculum with the duckbill closed and the handle in the upward position. Insert with care. Squeeze the handle and press down. You will hear a click to let you know it is locked open. To see yourself, hold the mirror between your legs and direct the light toward it. The light will reflect off the mirror into your vagina so that your cervix will pop into view. Enjoy the lush color, texture, odor, and shape of the cervix and vaginal walls. Take turns sharing your observations with the group. Admire the subtle variations and the fine differences in form. Track changes.

These are some of the ingredients making up the feminist protocol of vaginal self-exam in Los Angeles in the 1970s. This assemblage of commercially available devices, behavioral scripts, affective economies, and embodied subjects became feminist self help's iconic practice. Beyond a health care protocol, I want to argue, vaginal self-exam exemplified a historically particular and politically charged refashioning of *objectivity*. At stake in the protocol of vaginal self-exam was how to see and to create knowledge about health and bodies. In other words, vaginal self-exam made manifest the epistemological stakes—the politics of how-to-know—crafted into the biopolitical project of feminist self help. Moreover, feminist self-help practitioners argued that embodied ways of knowing produced better

---

Excerpted from "Immodest Witnessing, Affective Economies, and Objectivity," in *Seizing the means of reproduction: Entanglements of feminism, health, and technoscience*, Michelle Murphy, pp. 68–101.

knowledge. While historians of science tend to look to professionalized disciplines—physics, biology, astronomy, statistics—or increasingly to high-tech domains—such as genetics and nanotechnology—to historicize modes of objectivity, this chapter takes up the practice of vaginal self-exam to draw attention to how the history of objectivity in the late twentieth century was also crafted through politicized interruptions by nonprofessionals and, more specifically, by lay researchers who situated themselves as the embodied "subjects" and "objects" of technoscience. [. . .]

As an attempt to practice research as a political project that could tell better truths, feminist self help in the 1970s drew together an affective economy of technoscience that hoped to challenge dominant practices. Embracing instrumentality, feminist self help did not just seek to simply reveal a new truth about reproductive health, but offered new practices for interacting with, caring about, and managing reproduction—to seize the means of reproduction. Vaginal self-exam, as a protocol, explicitly attempted to operate outside of professional and profit-driven biomedicine, and hence grappled with the role of capitalism and authority in knowledge making by virtue of crafting alternative affective, embodied, and political, rather than economically productive, epistemic values.

[. . .] Vaginal self-exam, as generated in an affective economy of technoscience, worked to materialize bodies and researchers in new ways, on the one hand, and was caught between feminist counter-conduct and broader historical developments, on the other. As part of the history of objectivity, vaginal self-exam signaled the emergence of affectively charged practices as a core epistemic value. This value was both heralded by feminists committed to emotion and embodiment in knowledge making, and given monetized value in gendered labor and entrepreneurial technoscience at the end of the century.

Immodest Witnessing

[. . .] The protocol for vaginal self-exam was disseminated in an abundance of instructional images in slide shows, mimeographed handouts, films, pamphlets, and books. The Los Angeles Feminist Women's Health Center, in particular, took hundreds of photographs, some of cervixes, some of genitalia, others of the act of vaginal self-exam and other forms of appropriated biomedical labor. Not simply a straightforward set of written directions, these visual practices and materials offered a particular way of visually manifesting a protocol as a practice of seeing. [. . .] Moreover, I want to argue, such visual practices were constitutive of a reassembled status of the subject in objectivity. I will call this new subject-figure the *immodest witness*.

In tracking the figure of the immodest witness of vaginal self-exam in the visual productions of feminist self help, I am inspired by the "material-semiotic" figures in the scholarship of Donna Haraway, where she takes the Cyborg, the OncoMouse, or the Modest Witness as oppositional, and yet noninnocent, means to query technoscience. For Haraway, these are "performed images that can be inhabited" (1997, 11). What I am calling the *immodest witness* is likewise a complex oppositional and yet entangled subject-figure, incited into being not only in images, but also in practices, bodies, and affects. The visual tropes of vaginal self-exam functioned as procedural instructions, yet also as a generative reassembly of subjectivity and objectivity through embodiment. Starting with yourself in what you were studying, and highlighting your affective entanglements, were epistemic values that aspired to produce better, more accurate, knowledge. In addition, this vantage point promoted entanglements that offered a version of the scientist-subject as deeply responsible and implicated in her object of study. Simply put, who you were affected what you could know. For the immodest witness, subjectivity was not an abstract problem of seeing but a question of concrete and particular embodiment that promised a better—a more proximate and intimate—route to objectivity. [. . .]

Immodest witnessing [. . .] was explicitly both an object-making and subject-making process that elevated the layperson as expert in the particularities of herself. I use the word *immodest* here to draw attention to the project of laying bare the importance of the subject in knowledge making, and of challenging notions of chastity and modesty that prevented women from displaying, valuing, or studying the female reproductive body, or even marking the subject-figure of the scientist as sexed, and hence as a particular, not abstract, person. The visual practices of vaginal self-exam boldly announced the sexed embodiment of the laborer in knowledge production. For practitioners, the immodest witness was part of a tactic of "demystification" concerned with unmasking the craft of knowledge hidden by professionalism, thereby drawing attention to who was allowed to participate in the labor of science, revealing what had previously been obscured as actually the product of relations of power. [. . .]

At the same time, in their elevation of experience and sensation as epistemic virtues, radical feminists tended to consider their knowledge-making practices as a return to an empiricism associated with the scientific revolution: "The decision to emphasize our own feelings and experiences as women and to test all generalizations and reading we did by our own experience was actually the scientific method of research. We were in effect repeating the 17th century challenge of science to scholasticism: study nature, not books, and put all theories to the test of living practice and action" (Sarachild 1975, 145). In this way, the feelings and experiences of immodest witnessing became a primary passage point through which the validity of already existent knowledge—such as Marxist theory or biomedical descriptions—had to be tested.

The immodest witness was cleverly captured in the canonical self-help image of a woman examining herself with a mirror and a speculum. Unlike contemporaneous drawings of pelvic exams in gynecological textbooks—typically either a straight view into the vaginal canal, evoking the camera angles of pornography, or a cross section of disembodied organs with arms, legs, and head severed—images of the immodest witness put the viewer in the eyes of the woman examining herself. Our gaze is taken over our own pubis and into the mirror we are holding between our legs. In the mirror, the speculum guides our gaze to the cervix, yet the mirror as symbol of a transparent access to the world is resisted, for the illustration makes us aware of the mirror's frame and interpellates us into our own embodied gaze. The sex of the observer could not be missed.

Acts of women studying their sexed bodies through their bodies created a recursive circuit that joined the observer and the observed in a single gesture. This conjoining, first, rendered the body under observation an object of inquiry active in its own observation and, second, rendered the observer an embodied figure entangled with the object under study. Hands were an important trope in the figuration of the "möbius" agency of the immodest witness, intended to convey the use of the observer's senses, the generative lushness of the body itself, and the agency of the woman being examined. For example, hands figured prominently in the feminist self help project of crafting a "new definition" of the clitoris, which dramatically expanded the anatomical scope of the clitoris, as well as provided a sense of its detailed function. Illustrations of clitoral anatomy (drawn by Suzanne Gage, who developed much of the visual vocabulary for feminist self help) began with four drawings of fingers spreading the outer lips of a vagina, pulling back its hood, rolling the shaft, and squeezing the glans. [. . .]

The feminist self-help work on the clitoral study of 1978 was some of the most sexually charged research conducted, in which the process of orgasm itself was studied in a group (FFWHC 1991). [. . .] The revised clitoris was brought into lively and timely embodiment through pleasure, touch, and sight in both the registers of practice and representation. This affect-drenched research into genitals, enfolding the feeling object with the desiring subject who disidentified with norms, was distinctly a form of queered research. While introductions to vaginal exam were able to avoid questions of sexuality as the signature upturned speculum handle obscured the clitoris from the scope of introductory observation, "advanced groups" were much more likely to sexualize their affective economies of research. Advanced collectivities, for example, took up questions of ejaculation and lesbian health in intensely affective circumstances (Hornstein 1973). Thus, the immodest witness was also potentially an explicitly queered subject, who violated heteronormativity, not only by assuming the status of the scientist, but also by virtue of the affectively charged same-sex circuits of sensory observation of parts of the body deeply saturated with sexuality.

Affective Economies and the Not Uncommon

[. . .] Feminist self help did not use the term *affect*. It did, however, use the term *experience* as an emotional, sensual, and embodied value. Therefore, it is important to think through the question of affective economies in immodest witnessing by first considering the epistemic privilege with which feminist self help, as with radical feminism more generally, imbued "experience." Perhaps the most crucial axiom of consciousness raising and feminist research was that all knowledge production should begin with women's experiences. With feminist self help, however, experience was both the empirical material analyzed (the embodied experience of being a "woman") and the immediate encounter with one's body produced through vaginal self-examination (the experience of looking at oneself). At work in statements such as "I saw this," "I was there," "I felt that" uttered at feminist self help meetings was the assertion of a purported epistemic privilege gained from the immediacy of observing one's self. It shouldn't be surprising, then, that the movement's literature is loaded with the term *experience* and that its uses were both tangled and polyvalent. [. . .]

Though it may be tempting to take "experience" as a self-evident originary point of explanation—as that which explains, not that which needs to be explained—I want to attend to how claims to represent experience operate by taking as given and already constituted the identities of those whose experiences are being represented; whereas the task of the critical historian is to excavate the production of subjectivities though the ways the evidence of experience is imbued with an authentic primacy. In other words, the "evidence of experience" needs to be historicized. [. . .] The now worn phrase "the personal is political," coined by radical feminists in the late 1960s, was meant to signal the politicization of that which was previously held as personal, individual, and even trivial, not the personalization of politics into a private domain of self-improvement (Hanisch 2000). And like consciousness raising, the first lesson one learned in a self help meeting was that of commonality: "What you thought was peculiar to you was in fact shared by everyone" (Downer 1999). At the same time, the slogan "the personal is political" captured a danger within the method: the insight that social structures manifested themselves in what seemed like idiosyncratic personal events could be used to elevate quite historically and geographically particular insights into problematic universals. [. . .]

Vaginal self-exam, and feminist self help more generally, were yet another iteration and rearrangement of consciousness-raising. The role of the evidence of experience in vaginal self-exam differed from conventional consciousness raising in that it included the "immediate concrete" moment of examining one's own body (FFWHC 1981). [. . .] The evidence of experience joined together past reproductive, sexual, and medical events with the immediate moment of self-exam, which involved both the sensations of the speculum and the affective relations involved in the group, as well as the sensations of observation—what one saw, smelt, tasted, or felt. Immediacy was conveyed through rich sensory narratives: the feeling of pressure as a speculum clicked into place, the pinkish color of the cervix with or without reddish hues, the moisture or dryness of the vaginal canal, the sweet or musky smell of secretions, the look of the curly or toothy flesh of a hymen. A woman might even taste the sticky residue left on the speculum once it was removed. [. . .] Biological variation—idiosyncratic health histories and anatomical quirks—were the incidental experiences to be gathered through a fine-grain corporeal attention. When anatomical variations were collected in the self help clinic, feminist self helpers pointed to a shared reproductive body underneath—"below the waist and above the knees"—but this shared domain was nonetheless lively with variation (Gage 1999). Their intimate examination of reproductive variation was not primarily a search for ill health; in contrast, it was an effort to remove reproduction from its association with pathology—"taking the routine into our own hands"—and revaluing embodiments in terms of, not in spite of, individual biological deviations.

Further, instead of a straightforward search for the undergirding common that characterized much consciousness raising, vaginal self-exam was crafted as a means to recognize that the "irregular is not uncommon." This double negative of "not uncommon" is crucially different from "common." Vaginal self-exam, as well as the research in "advanced groups," sought to study the not uncommon as routine anatomical variation, as ordinary secretions, and as ubiquitous infections so that these phenomena could be depathologized and seen as more appropriate to "home care" by women themselves than to medical care by doctors.

Attending to the not uncommon of reproductive health was likened to oral health practices—in terms of teeth brushing, gargling, self-inspection of the mouth. Both were practices done outside of medicine. Authority to judge one's own vagina and thus "demystify" reproductive anatomy was made analogous to the unexceptional act of examining one's own mouth. Unlike an organ—a technical term that might spring to mind when hearing the gynecological term "internal exam"—the mouth was an accessible cavity laypeople regularly inspected, took care of, and treated. Both were "open to the outside" with mucous membrane linings; neither were a sterile environment (FFWHC 1981, 24). Many things are put in the mouth, and so too with the vagina: fingers, penises, tampons, spermicidal foams and jellies, diaphragms, douches. And other things came out, not least of which were babies. Thus, according to self help protocol, a woman should feel licensed to have the same access and relationship to her vagina as she does with her mouth. [. . .]

At the same time, the "not uncommon" was also a valuation of variation itself. Variation was its own epistemic virtue and, moreover, variation gave the evidence of experience a particular form, one which was concerned with searching for and positively appreciating idiosyncrasies. To this end, over the course of the decade the Federation of Feminist Women's Health Centers (FFWHC) took hundreds of pictures of genitals and cervixes, recording a lush field of individual variety. In this way, so called not uncommon problems were refused the label of pathology or deviance, and instead were heralded as unexceptional variations that non-professionals could recognize, monitor, and manage. [. . .]

By attending to the personalized and interindividual difference within a group as the not uncommon, vaginal self-exam offered a mode of collecting data that called into being what I will call an *ontological collectivity*—a materialization of a continuous field of difference rather than a fixed form. That is, interindividual differences were not judged by an abstract norm, or in terms of a fixed fact; rather, they were assembled into a collectivity of living variation both between bodies and within any given body. Thus, for feminist self help, the category "woman" was not absolutely unitary; instead it was a living ontological collectivity across bodies and over time. [. . .]

The visual vocabulary of feminist self-help images captured this epistemic value of variation. Most images were not only of embodied women; they were of the body of a particular woman, who might sport a pair of glasses, have scraggly pubic hair, or slouch (Gage 1999). The women represented in these images were clearly raced, diverse, and individual. Specificity mattered, and immodest witnessing sought to carefully attend to the specificities of the individual body as well as the variations between bodies, corralling these variegated bodily expressions into the ambit of the figure of the "well-woman" (For more on the politics of the term well woman, see Murphy 2012b). Thus the immodest witness was the exemplar of a quite remarkable reassembly of objectivity that altered, one, the status of the subject (as particular, embodied, and affectively entangled with its object of study) and, two, the epistemic values of observation (valuing variation, sensation, and emotion individually and collectively).

At the same time, attention to the not uncommon asked participants to recognize themselves in each other according to the protocols of consciousness-raising group work. In fashioning this web of mutual summoning, practitioners were not simply discerning common patterns; they were evoking each other as politicized and connected subjects. Speaking across a circle, women recognized each other as embodied agents capable of truth claims. "Responding with recognition" thus asked women to valorize one another as highly individualized truth-tellers and to align as part of a politically charged cohort of "women," a more abstract commonality. Thus, the importance of individuality to the ontological collectivity created by vaginal self-exam was in tension with the necessity of invoking woman as a universalizable sex. [. . .]

The immodest witness, therefore, was formed through contradictions. Practitioners were simultaneously hailed as representatives of a politicized class (women) and as singular individuals. They were simultaneously implicated agents responsible to that which they studied, and an object of inquiry that spoke for itself. Immodest witnessing was structured by this tension of interplay between women as variegated individuals, as members of a common class, and as participants in a research group collectivity. The assumed common sex of participants excluded other ways of marking difference and drawing together collective life. Moreover, the values of individualized variation and comparison among peers was premised on the bracketing off, and even the erasure of, the complex circumstances that placed some women as agents over the fate of others. The assumption that women were invested in the fate of each other simply by virtue of also being women belied the contradictory ways women were riven and bound by uneven biopolitical topologies of late twentieth-century America (for a detailed version of this argument, see Murphy 2012a). While the figure of the immodest witness performed a radical implosion of the subject/object in observations, making visible the embodied and affective subject in the production of knowledge, the immodest witness simultaneously tended to foreground sexed and individualized embodiment at the expense of a more complexly situated, interlocking, and thus more complicit, map of subject making in knowledge production. [. . .]

In sum, affective entanglements in practices of vaginal self-exam circulated in multiple dimensions. In the moment of immodest witnessing, knowing was an embodied, sensory, and recursive act that situated subjects as particular. In the collective project of feminist self help, affective entanglements formed a moral economy of affirmation—of the happiness of knowing oneself through bonding and of recognition of oneself in others as a politicizable collectivity. At the same time, objectivity was reassembled as a project of self-knowing only possible in politically and affectively charged relations with other subjects.

Immodest witnessing, however, can also be examined through an expanded sense of "affective entanglements." The epistemic value granted individualizing and yet bonding affective knowing was, in the 1970s, attached to a larger political economy that helped to set its conditions of possibility. On the one hand, affective entanglements were a form of counter-conduct reacting to practices of dispassionate, professionalized, patronizing, and even coercive scientific authority. On the other hand, they exemplified tendencies in contemporary American culture that called on subjects to release and express their feelings and desires as political, therapeutic, and entrepreneurial acts. While calls to name and fulfill desires through lifestyle consumption and entrepreneurial vitality were deeply antagonistic to radical feminism, such injunctions were nonetheless part of feminist self help's world, and implicated with their effort to elevate affect as a means to find emancipation (see, for example, Rofel 2007). In other words, feminist self help was animated in broader circulations of affect that went beyond their own practices. [. . .][1]

Ontological Collectivities, Liveliness, and Biovalue

[. . .] It was in the advanced research groups of feminist self-help that the most intensive efforts to rematerialize reproduction as a lush liveliness took place. [. . .] In self help clinics, participants were encouraged to take daily observations on their own, perhaps including a quick sketch of what they saw in a journal or calendar that they could later puzzle over with the group. These repetitious chronological traces could be assembled into a portrait of minute change over time, further expanding the topography of variation into the dimension of temporal change. Rather than comparing themselves to an abstract, universalized norm (as one might find in a medical textbook), in using the technique of vaginal self-exam they relied on comparisons within small groups of women and with each woman's own changes over time. This schooled attention to slight variations in anatomical detail over time produced a sense of lush, changing variety through which the feminist self help movement sought to remap the anatomical terms of healthfulness. Healthfulness here was a disidentification from abstract norms in favor of shifting dynamic forms.

This rearticulation of healthfulness as a lush garden of shifting anatomical diversity extended beyond macroscopic features to include microorganisms ubiquitously present in vaginas that could cause common and minor, though sometimes recalcitrant, infections. This microscopic variation was dubbed the "ecology of the vagina" (FFWHC 1981, 24–25). While viewing a drop of vaginal secretions with a microscope, self helpers taught themselves to see "sloughed-off cells from the vaginal wall, a few yeast plants, lots of bacteria and sometimes even a few one-celled animals, trichomonads" (FFWHC 1981, 24). If a woman was menstruating, she would see red blood cells. If she had recently had heterosexual intercourse, she might see sperm. As with their gross observations, self helpers were concerned with noticing how the exact constituency of a vaginal ecology would change over time, often in synchrony with the changing pH of the menstrual cycle.

This sense of a vaginal ecology was aided by microscopy. A manual of procedures in a Well Woman clinic, called the *Black Book* (written to defend against accusations of practicing medicine without a license), explained that using a microscope "is simply an aid to better eyesight" (FWHC 1976). [. . .] Though strategically represented as a simple magnification of eyesight, the ability to perceive a wet mount slide as a vaginal ecology was by necessity a learned technique of observation that called for assembling details into a relationship of changing diversity. Looking in a microscope was not a neutral gaze taking in a self-evident world; it was a repeated act made sense of by a politicalized apparatus (the self help clinic) that re-represented entities already codified by conventional gynecology. [. . .] The practice of vaginal self-exam not only schooled women in fine-grained sensory observations, it simultaneously refigured their object of study—vaginal ecologies—in a way that authorized the very act of intimate frequent personal observation.

The elaborate daily attention to minute changes of one's vagina and cervix reached its pinnacle in the Menstrual Cycle Study of 1975, also undertaken as part of the Federation book project.[2] Every morning, during a full menstrual cycle, a cadre of nine women gathered to make thirty-six time-consuming observations about their own bodies. [. . .]

What was rendered perceptible through this elaborate and tedious collecting and cataloguing of detail? The study did not conclude with a summary description of a menstrual cycle. Nor did it identify a series of markers that identified distinct stages. Instead, the study concluded what it was designed to perceive—precisely the converse: women do not match an abstracted cycle, and healthfulness cannot be accurately measured through a once-a-year marker like a Pap smear. [. . .] What a "normal" menstrual cycle looked like, so they argued, could only be determined by studying "each woman's cycle within the context of the cycle itself as opposed to comparing to a norm," thereby "redefining and individualizing the concept of 'normal' for women" (FWHC 1975). Since doctors relied on annual visits and did not have the time to make such painstaking daily observations on each patient, women occupied a privileged position for understanding this complexity. Using odor as an intimate organizing affective entry point could, for example, produce a disorienting anatomical portrait unlike that found in any medical textbook. Feminist self help strongly encouraged women to use smell, and not just eyesight, to track their vaginal ecology. [. . .]

Materializing reproductive embodiment as a highly valued, dynamic, and generative domain of variation and affect, through practices explicitly considered radical, nonetheless had a synergistic relationship with new modes of valuing living-being, particularly in research with biotechnologies indebted to the reproductive sciences. While feminists of the 1970s were more likely to characterize medicine as treating women's bodies like machines in industrial assembly lines, and while this was certainly a dominant organizational motif for apprehending the body, the 1970s was also a moment of emerging biotechnology, particularly in California, and the beginning of a new era of clinical reproductive medicine concerned with the technical choreography of fertility at micrological levels (see Davis-Floyd 2003; Martin 1991). Through the arrival of reproductive medicine, the 1970s was an important decade for the refiguration of what the feminist technoscience scholar Catherine Waldby calls *biovalue*. Biovalue names the "yield of vitality produced by the biotechnical reformulation of living processes" (2002, 310). [. . .] As I argued above, calls to express individualized desires not linked to mass culture could enable new modes of hitching affect to capital in marketing. In a similar way, the oppositional ontological politics of heralding the ways living-being can synergistically generate difference and affect reverberate with capitalism's requirement for exuberant territories through which to implant and recycle value.

Topologies of Situated Feminist Technoscience

[. . .] As a situated knowledge, then, the broad lesson to be learned from historicizing vaginal self-exam is that the "how" of knowing is as much a question of the promise and limits of affectively charged counter-conduct, as much a question of subjectivation in noninnocent economies, of entangled reassemblies and appropriations, and of marked and unmarked labor as it is a historical episode in the history of objectivity. Rather than an odd marginal practice only of interest to feminists, vaginal self-exam announces a particular reassembly of objectivity in the late twentieth century, with the generative features of counter-conduct, affect, and biovalue at its heart. [. . .]

Feminist self help, with vaginal self-exam as its iconic protocol, was one of the most sustained efforts to practice science as feminism. It was also a reassembly of objectivity that reverberates in participatory methodologies of many political stripes today. What was the fate of vaginal self-exam? Rarely practiced today, vaginal self-exam declined both for reasons from within and without. As the Reagan era unfolded [. . .] militant antiabortion activists besieged Feminist Women's Health Centers. The day-to-day harassment and the imminent threat of violence created a "siege mentality" within the centers' walls (Hasper, Heckert, and Schnitger 1999). The incredible amount of energy, emotional and physical, that went into escorting women through blockades, into clinic security, into court cases, into finding doctors willing to work under the threat of violence, and into rebuilding destroyed clinics drastically redirected the labor of the feminist self help movement.

From within the women's health movement, their very success deflated the consciousness-raising power the vaginal self-exam had enjoyed in the 1970s. A new moral economy of health care arose—calling for the well-educated, well-informed, self-knowing patient to be prepared to advocate for herself as a consumer within corporate medical institutions. Put simply, in the last thirty years the status of white, middle-class women as patients has dramatically changed, and thus so too did the biggest constituency that the affective economy of vaginal self-exam had appealed to. While the practice today is rare, the assembly of affective engagement, particular embodiment, injunctions to empowerment, intimate attention, collective labor, and generative living-being coils forward.

Notes

1 *Editors' note:* See unabridged chapter for discussion of "antagonistic and yet enabling relations" between feminist affective projects and economic valuations of affect.

2 The women who participated in this study were Suzann Gage, Carol Downer, Karen Grant, Lynn Heidelberg, Kathy Hodge, Frances Hornstein, Margo Miller, Sylvia Morales, and Lorraine Rothman.

Works Cited

1973. *Lesbian Health Activism: The First Wave. Feminist Writings from the Early Lesbian Health Movement.* Los Angeles, CA: Feminist Health Press.

FFWHC, Federation of Feminist Women's Health Centers. 1981. *How to Stay Out of the Gynecologist's Office.* Hollywood, CA: Women to Women Publications.

———. 1991. *A New View of a Women's Body*: Feminist Health Press.

FWHC, Feminist Women's Health Center. 1975. *Self-Help Study: Observing Changes in the Menstrual Cycle.* Los Angeles, CA: Feminist Women's Health Center.

———. 1976. *Well Woman Health Care in Woman Controlled Clinics.* Los Angeles, CA: Feminist Women's Health Center.

Davis-Floyd, Robbie. 2003. *Birth as an American Rite of Passage*: University of California Press.

Downer, Carol. 1999. "Interview." October 24, 1999.

Gage, Suzanne. 1999. "Interview." October 25, 1999.

Hanisch, Carol. 2000. "The Personal Is Political." In *Radical Feminism: A Documentary Reader*, edited by Barbara A. Crow, 113–116. New York: NYU Press.

Haraway, Donna. 1997. *Modest_Witness@Second_Millennium.FemaleMan©_ Meets_ OncoMouseĭ: Feminism and Technoscience*: Psychology Press.

Hasper, Heckert, and Schnitger. 1999. "Interview." November 1999.

Hornstein, Frances. 1973. *Lesbian Health Care.* Los Angeles, CA: Feminist Women's Health Centers.

Martin, Emily. 1991. "The Egg and the Sperm: How Science has Constructed a Romance Based on Stereotypical Male-Female Roles." *Signs* 16 (3): 485–501.

Murphy, Michelle. 2012a. "Assembling Protocol Feminism." In *Seizing the Means of Reproduction: Entanglements of Feminism, Health, and Technoscience*, 25–67. Durham: Duke University Press.

———. 2012b. “Pap Smears, Cervical Cancer, and Scales.” In *Seizing the Means of Reproduction: Entanglements of Feminism, Health, and Technoscience*, 102–149. Duke University Press.

Rofel, Lisa. 2007. *Desiring China: Experiments in Neoliberalism, Sexuality, and Public Culture*: Duke University Press.

Sarachild, Kathie. 1975. “Consciousness-Raising: A Radical Weapon.” In *Redstockings, Feminist Revolution*, 144–50. New York: Random House.

Waldby, Catherine. 2002. “Stem Cells, Tissue Cultures and the Production of Biovalue.” *Health:* 6: 305–323. doi: 10.1177/136345930200600304.

Pussy Ballistics and the Queer Appeal of Peristalsis, or Belly Dancing with Margaret Cho

RACHEL LEE

> [It is through] "the peristaltic movements of the viscera, the mitosis of cells, the electrical activity that plays across synapse, the itinerary of a virus" . . . that feminism, in our current context, may gain its most effective political purchase on biology.
>
> —ELIZABETH WILSON, QUOTING VICKY KIRBY

> How powerful women are . . . we bring forth life. I was so . . . amazed [witnessing] my [friend's delivery] because at that point, she was not just a woman, not just a mother. She was creation. She was life. She was god. . . . And, as I looked into her eyes [blows hard into the mike] . . . HER PUSSY EXPLODED!
>
> —MARGARET CHO

THIS RUMINATION ON THE VISCERA IS INSPIRED BY AND CONTINUES my long-standing engagement with the work of the Korean American stand-up comedian Margaret Cho, self-nominated "fag hag," spokesperson on the tyranny of skinniness, and commentator on race relations in the contemporary United States (Lee 2004). Well-known for her vagina

---

Rachel Lee, "Pussy Ballistics and the Queer Appeal of Peristalsis, or Belly Dancing with Margaret Cho," in *GLQ:A Journal of Lesbian and Gay Studies*, Volume 20, no. 4, pp. 491–520.

monologues—as in her uproarious retelling of a presurgical experience with a nurse-attendant bent on "waaaaarshing her vagina"—Cho returns to this and other intimate bodily zones in her 2003 concert *Cho Revolution*, but this time emphasizing the ballistic activity of the tissues themselves—the network of erogenous, incubational, and vestibular organs making up the so-called reproductive tract (Cho 2004). While the viscera conventionally refer to the entrails—the internal organs of the abdominal and thoracic cavities—my rumination on the visceral qualities of Cho's work begins, as she does herself, with the vaginal or "pussy" explosions of her 2003 live performance and moves from there to her engagements with the belly.

In the comic bit from which the second epigraph is drawn, Cho vacillates between horror and clinical calm at the sight of the birthing mother's exploding pussy. At first registering disgust at the disintegrated boundaries of her friend's vagina, Cho switches roles, playing the nurse-attendant nonplussed by the displacement of the innards. She continues, "The nurse was running around collecting pieces of her pussy in a basket." Cho pantomimes collecting debris, then lunges low to grab a hard-to-reach chunk from under a stool. "They had to sew it back together," she says, while weaving her hand in big figure eights, contouring perineal suturing as embroidery work. She then finishes off the knot by tearing the thread between her teeth. The scandal lies in the nurse's nonchalance: her ho-hum attitude toward fleshy proximity (her mouth next to the birthing mother's perineum) and cross-bodily confusion (the nurse plucks "pieces of pussy" out of her hair) provokes nervous titters and thigh-slapping howls from the audience.

Returning to her role as witnessing friend, Cho the actress concludes the segment by announcing her decision never to do the same: "I'm not a breeder. I have no maternal instincts whatsoever. I am barren. I am bone dry. When I see children, I feel nothing. [Long pause] I ovulate sand." (While Cho biologizes her progeny-less state for comic effect, her earlier written memoir states, quite to the contrary, that after becoming pregnant in the 1990s, she electively terminated the pregnancy (Cho 2001).) Cho refuses the imperative to breed, rejecting a heteronormative futurity predicated on, as Lee Edelman puts it, fighting for the child (2004). Indeed, one might read Cho's rejection of the procreative imperative as many things: solidarity with queerness, an exercising of feminist choice (the right to abortion), and also self-doubt over her own capacity to take up the mantle of Korean motherhood. At its most basic level, however, the sequence stages an anxiety around being biological—specifically, around the morphological, hormonal, and visceral changes induced by pregnancy and birth routinized as women's natural work and deepest fulfillment. At the same time, in claiming to "ovulate sand," Cho *biologizes* her refusal of maternity rather than appeal to feminism's discrediting of the idea that a woman's worth lies in her womb.[1]

Much later in the show, the comedian riffs on this earlier witnessing of a "pussy explod[ing]." Setting up the joke, Cho speaks of visiting Bangkok where hired doormen tout the live acts inside:

> "Pussy eat banana!"
> [Cho mimes a look of disbelief.]
> "Nooo, thank you . . . not in the mood for dinner theater."
> [The barker continues]
> "Pussy play Ping-Pong!"
> [Cho hesitates, inching toward these promised entertainments]
> ". . . Pussy . . . write letter!"

This "pussy" does not scatter into parts and splatter onlookers' bodies, but rather an incessant hawking of its capacities keeps it in contagious circulation. As I show, with this second reference to an explosive pussy—that of the Ping-Pong-propelling Thai sex worker—Cho initiates a leitmotif where she randomly barks out commercial "deals" and practical "tricks" that pussy will do for you: "Pussy change oil every 3,000 mile!" Cho and her audience are repeatedly terrorized by these pussy blasts across the length of her show. Put another way, the worry parlayed via the specifically Thai vagina takes the contours not of pussy literally exploding into fleshy fragments but of *vocal* explosions that spectacularly document that pussy's hyperanimate possibilities. What this essay's title calls pussy ballistics refers at once to (1) the parturition characterizing the privileged and procreative good mother of the global North, (2) the contrastive gratuitous explosions from Asian sex workers' vaginas (e.g., of Ping-Pong balls), and both of these to (3) the gut(teral) explosions from the mouth that sell Asian pussy and terrorize the audience with their surprise regurgitations.

Examining the way these pussy bombshells erupt strategically across Cho's concert, this essay explores how Cho's stylistic repertoire mounts a critique of US empire that focuses on the Asian sex worker as logistical support for the military troops of commodity capitalism. In attending to this formal aspect of the concert's overall structure, I develop a theory of pussy ballistics, proposing its value as a mode of pursuing feminist, queer, and racial studies critiques precisely through its roundabout technique—its refusing the straight path going directly to the point. As pussy ballistics incite rhythmic waves of (commercial) terrorism that first register below the belt as belly-shaking laughter, the felt actions of the viscera intrude into the cognitive registers where apparent rationality does not rule as much as (imperialist) disavowal and contradiction. To put it another way, Cho's lobbed pussy shout-outs wash over her audience like a peristaltic wave.

Peristalsis, both a propulsive and radial action that moves materials into and out of a cavity or through a tube in a wavelike motion, characterizes the uterine-cervical tract; the walls of the esophagus, stomach, and intestines; and the linings of the ureters, bladder, testes, and vas deferens. It describes as well the process whereby nonhuman organisms such as annelid worms locomote. Though Cho herself never names the underlying muscular agency of gut and uterus as peristaltic, she amply enacts and vocally performs this kinesthetics as a motility produced by an unconscious flesh-coordination shared across regions conventionally considered distinct zones of the body (e.g., the reproductive and the digestive systems, the uterus and the mouth). *Cho Revolution*, I argue, tackles so-called consequential matters of war, empire, occupation, and militarism by way of imitating both the pyrotechnics of high-tech armaments and the lowly contractions of the smooth muscles lining, yes, the viscera but other bodily zones as well. [. . .]

As suggested by my first epigraph, my methodological endeavor in this essay is part of an ongoing development in embodiment studies and feminist/queer (hereafter "femiqueer") theory to resituate the biological sciences as a resource for feminist and queer thinking (Ha 2011; Hird 2012, 213–37; Margulis, Sagan, and Korein 1986; Roy 2008, 134–57; Roughgarden 2004, 2009; Subramaniam 2014). Joining a science and technology studies approach to a performance and literary studies method attentive to "form"—to aesthetics-kinesthetics—my project seeks a more precise lexicon for how artists, like Cho, make queer (or homo) any pure/puritan distinctions between organ systems by re-membering their visceral "amphimixis."[2] Doing so broadens the reach of queer approaches and contours the relevance of race and geopolitics to domestic matters of "health, reproduction, and well-being"—those focalized arenas of contemporary biopower. Across this essay, I propose peristaltic action as a rejoinder to a feminist politics premised on moral virtue and ask whether peristalsis provides a positive model for a queer onto-epistemological method that acknowledges political critique working most honestly when accepting the embodied viscera's always sullied and "bottom" positions. [. . .]

Babies in Mouths and Dogs in Pots

Cho chooses not to be a "breeder," opting *out*, it would seem, from the utter messiness of birth and refusing a heteronormative futurity premised on fighting for the child. Yet others would still enroll her in the caretaking—the assuring of a secure environment—for their children, as dramatized by the conclusion to the sequence with which I opened this essay. After remarking that her friend (whose labor and delivery Cho witnessed) has turned out to be an excellent mother, the comedian pantomimes that same friend thrusting her baby girl into Cho's arms despite the latter's gesticulating aversion. The sequence ends with Cho, finally resigned to taking the child in her arms, gulping her down: "So I just ate her." By swallowing whole this avatar of a future that must be defended against alien risks (including SARS, sex workers, and same-sex desire), Cho eats this non-nourishing item just to get rid of it.

That reproduction encompasses more than gametes (i.e., genetic materials) but also nutrition (breastfeeding, tactile stimuli, adequate caring touches, and so forth) is re-membered perversely in this episode of the adult gulping down the child rather than the infant gulping from the proverbial teat.[3] Significantly, across Cho's corpus, stories of food and eating repeatedly provide the mise-en-scène in which Cho is surveilled both racially and misogynistically (Lee 2004). The most obvious ways in which the comedian presents herself as subject to an intense anatomo-political regulation is through the demands from media networks. Cho recounts that, after her screen test for a prime-time situation comedy, *All-American Girl*, in which she would eventually star, the American Broadcasting

Corporation (ABC) set to air the pilot voiced concern over her size, hiring a personal trainer and dietician to resculpt her body because "clearly I couldn't be trusted to make my own fatty choices."[4] The studio never expressly says that it is the Korean-ness of her visage that strikes studio execs as uncomely, but carefully uses "deracializing racialist" euphemisms, to use Khiara Bridges's (2011) terminology—such as concern over "the fullness of [Cho's] face"—that target her physiognomic non-normativity in a pastoral mode of concern. When she later notes that her canceled situation comedy has been replaced by a show starring a hefty white male comedian, namely Drew Carey, "because he's so skinny," Cho pointedly underscores the gendered and racialized aspects of the meager food provisioning required as a condition of *her* twenty-first century employment in the entertainment industry, but not *his*.

Here and elsewhere Cho telescopes encounters at the table (issues of cuisine, nutrition, and eating disorders) as the site of her racialization, a racialization taking place as well in "everyday" encounters with strangers, rather than strictly in more formal repertoires of biopolitical governance (e.g., public health awareness campaigns, best practices of clinical care, protocols of drug trials). She flatly states that "something racial happens to me every day," with *IOIW* devoted partly to limning how "race" is spoken without literal reference to skin color, eye shape, nose width, and hair texture and color but is nonetheless operative. In one of the best examples of turn-of-the-twenty-first-century racialization theory pursued in and as a joke, the comedian ends that concert by mimicking the contours of addiction and recovery narratives. After enduring studio executives' and her agent's attempts to retire her outspoken comments on racial matters (indeed, her sitcom is canceled), Cho spirals downward into alcohol and drug addiction, laxative abuse, bad boyfriends, a brush with the casting couch, and an accidental pregnancy and elective abortion—all consequent on the trauma of Hollywood racism. Recuperating from these multiple symptoms of psychic and somatic trauma, Cho adopts an animal companion: she picks the saddest-looking dog at the shelter, saying "that dog is me," and nurtures him back to health:

> I stopped drinking and I got better. And as I got better, my dog got better. He's the greatest and we walk everywhere together. And people talk to you a lot more if you have a dog. Because I was walking the dog, and

> this homeless guy jumped out and said, "That dawg gonna wind up in a pot o' rice . . ." [Long beat, as Cho mugs an aristocratic shock at the impropriety, hand to the oval formed by her lips (*o*) held until the laughter dies down.]
>
> "And he probably wouldn't have said that if I was by myself."

The tagline, here, strictly adheres to the race-blindness suggested by one of Cho's former managers who encourages her toward a more universal humor because, as he puts it, "the Asian thing puts people off." The humor of the above joke, in other words, turns on those who cannot "get" either the homeless guy's remark (implying that Cho is going to eat her dog) or Cho's silent way of poking fun at pronouncements intimating the arrival of a postracial world. We might frame Cho's later comic bit in *Revolution* (three years after the above sequence), about eating her friend's baby, as performing an insider's citation and "terrorist" enactment of the racist epithet about Asians as those who eat (up) their cuddly domestic companions rather than caretake them properly (Muñoz 1999, 108). Most saliently, Cho's mode of joking about racism expressed as dominant culture's disgust toward "foreign" cuisine occurs via unspoken channels: she renders silently meaningful how racialization happens by way of deracialized racialist demands made on the marked, but not the unmarked, person. Racial oppression slyly punctuates this final section of *IOIW*, performed in/as the avoidance of race.

As noted earlier, *Revolution* continues Cho's focus on racial issues at the table and her struggle with weight, but oddly while the performer's visage and figure have visibly narrowed. In interviews, Cho attributes her new slenderness to her learning to belly dance and to her ceasing the hypersurveillance of her food intake (Moncur 2005; Cho 2005). Pathologies linked to eating and consumption are rediagnosed (reenrolled to wider causation beyond the individual "patient") through Cho's "pussy" shout-outs. Ushered in by the supposed failed (screen) test of her Asian bodily fitness ("the fullness of her face"), Hollywood clinicians including a trainer and a dietician treat Cho's facial fullness as a symbol of her deficient moral character in not refraining from overeating. From that narrow locus of Cho's person as the site of intervention, pathologies linked to eating and consumption are reproposed by Cho as material manifestations of circulating impersonal affects saturating the environment through promotions for tasty edibles ("Pussy come with ranch, thousand island, blue cheese, vinaigrette"; "Pussy come with fruit on the bottom"). Cho figures the tyranny of body image and eating disorders as very much an issue of hers—the possession of a "cured" or optimal slim figure changing nothing with respect to her political advocacy—and an issue of everyone else washed over by this promotional network.

Standing before her audience in the physiognomic container of a slim and petite Korean American beauty, Cho nonetheless stresses this transfiguration into monocultural optimal skinniness as utterly contaminating. She recounts that, while on an all-persimmon diet, driving her car in LA traffic, she finds herself compromised:

> I was driving in my car here in Los Angeles about four o'clock in the afternoon . . . kind of rocking out "Holiday" . . . [hums a few bars from the Madonna hit] . . . and I realized, I am going to SHIT RIGHT NOW!
>
> [mouth agape, horrified for close to a minute]
>
> [in a diminutive voice] And it caught me off guard . . .
>
> . . . because normally you have a good twenty minutes. There is this window of opportunity where you know to look for a Barnes and Noble or some kind of equivalent book-music superstore. We all take that for granted. But I did not have that luxury.

A ten-minute sequence ensues in which Cho speaks of bargaining with God, trying to release just a little, and then, as she flops her body to the ground, fists pounding the floor, she screams, "AND IT ALL CAME OUT!!" Next, she verbally evokes the squishy warmth of fecal matter turning cold, eliciting an audible groan from the audience. Cho slides farther down, mimicking deliquescence itself. Cho editorializes, "because I thought I was fat, ugly, because I believed those media images . . . and thought I was just gross. . . . I was now paying the price, by sitting in a pppwwhhhhoool [Cho vocalizes the word *pool* as if it were trisyllabic] of my own shit." In this tidy moral, the comedian suggests a straight chain of events: bad representations in the media upholding superskinniness leads to a woman wallowing in her own abjection. However, this sequence needs to be read in light of the earlier explosive riffs and later ones to follow. Instead of becoming a sanctified mother (with an exploding pussy as the proper feminine channel for extruding and gifting to society a part of one's body),

Cho spews gifts from her bottom orifices and makes her audience revel for a moment in those evacuations (parturitions).

Cho undercuts her tidy moral that would localize the object of this lengthy sequence's critical lesson to solely *her own* consumption issues. From her literal abject slide onto the stage floor, Cho metaphorically picks herself up (though she physically stays seated), by using talk and publicity as other channels of circulation that take over when she is immobilized in a pppwwhhhhoool of her own shit: "The only thing left to do at that point, was to call people," she mimes the flipping open of her mobile device. "'You better call me right back, because you are not going to believe what I just did. I'm in my car right now. I just shit my pants. And I'm coming over.'" Satellite technology provides the infrastructure of this cell phone circuitry. Empiric military residue, in short, "comes over" as fecal ooze, suggesting insecurity at the very core of a satellite-assisted will to total surveillance, even and especially when in the name of more and better control. Cho stresses the gargantuan efforts to contain, indeed, squeeze the life out of her Asian form, resulting in the contaminating return (the "coming over" en masse) of the designated "surplus populations" birthed from the hot wars and occupations of US nonterritorial empire and market penetration ("we [Asian immigrants] are here because you were there").

Cho's pussy explosions are only figural fissioning forces. She does not spread radioactive fallout in her virtual bombings; but what is not strictly figural is the radiating contagion—the lateral, sideways spread of her explosions. Laughter—the rhythmic, gut-shaking plosive of "har har" punctuated by gulps of air to feed more frequent shaking—also moves the abdomen in waves that resemble reverse peristalsis (vomiting). As the playwright Suzan-Lori Parks quips, "Laughter is very powerful—it's not a way of escaping anything but a way of arriving on the scene. Think about laughter and what happens to your body—it's almost the same thing that happens to you when you throw up" (1995, 6–18). To be sure, the laughter Cho provokes in her audience does not enact a literal reverse-peristaltic action; nevertheless, I call attention to the consilience between the visible belly-shaking kinetics of laughter and the invisible underbelly (intestinal) radial waves of peristalsis partly to follow her concert's own stress on the distributed agency and cognitions of the organs themselves—on zones of the body beyond the central command and control of the brain. Cho's audience, too, may want to let just a little out, but these hiccoughing intestinal waves have a rhythmic agency of their own, as anyone overcome by laughter knows firsthand. (And woe betide the concertgoer with any pregnancy—or otherwise—induced "incontinence.") Through speaking about but also inciting bodily leakage, Cho reminds us of her and our repertoires of belly dancing. She puts our brains back into our guts (as tactile, muscular, respiratory knowledge), and then has us laugh alongside the something mechanical (the militarized, consumer product edifices) encrusted on our wormlike gastrointestinal materiality (that paper over our diapered pasts and diapered future). Perhaps dwelling in our shit (as enacted in Cho's persimmon episode)—modeling ourselves more akin to worms and snails, babies and the aged—would have us understand better the consequences of militarism, of our culture devoted to death and dying as life's improvement and securitization.

With no pure, unsoiled platform to grab onto, Cho does not extricate herself from the excrement but only circulates news of it by phone. In effect, she refuses to respect the boundaries between high-tech GI ("government issue") equipment, as in the satellite technology of star wars, and the trivial emissions of that other GI (gastrointestinal) tract. The comedian draws out attention to bodily amphimixes that underlie the tissues of our psychic (thinking) and somatic (metabolic maintenance) systems as contoured by the feminist critic of the psychological field Elizabeth Wilson. In her highly suggestive account of the enteric nervous system and the distributed quality of visceral "thinking," Wilson speaks of the motive capacity or "psyche" of our biological parts distributed beyond the brain, claiming that "temper, like digestion, is one of the events to which enteric substrata are naturally (originally) inclined" (2004, 85). A digestive disease specialist has recently confirmed that signals from the brain not only affect the gastrointestinal tract but that signals from the human intestine are sent to the brain (Champeau 2013).

Key to Wilson's specific crediting of how the enteric nervous system and "gut" not only digest but ruminate (think, register dissent) is the biological notion of amphimixes:

> The rectum communicates its retentiveness to the bladder; the bladder communicates its liberality to the rectum. Without such inter-organ exchange, the bowel would become hopelessly constipated and the

> urinary tract incontinent. Amphimixes . . . is the very means by which these organs are able to function naturally at all. . . . Various organs of ingestion, expulsion, sensation and expression are borrowing from one another. (Champeau 2013, 81)

While Wilson, here, associates "amphimixes" with the expressive borrowing of one organ system (bowel) from another (bladder), we can think of amphimixes as referring to a leakiness or continuum of cellular memory among what later become distinct organs. Wilson's concrete illustration of amphimixes in this latter ontogenetic aspect is the soft tissue at the back of the throat—the fauces. This antechamber or protopharyngeal area is the "embryological source of several important structures in vertebrates . . . the breathing apparatus (gill pouches of fish and lungs of land animals) arises in this area. . . . In humans, the pharynx is particularly important as an instrument of speech." (Ferenczi, quoted in Wilson 2004, 80). It is also the site of the gag reflex, the aperture where mouth is connected to esophagus, and the meeting place of mouth, nasal passages, and ears: "The back of the throat is a local switch point between different organic capacities (ingestion, breathing, vocalizing, hearing, smelling) and different ontogenetic and phylogenetic impulses. . . . The fauces is a site where the communication between organs may readily become manifest" (Wilson 2004, 80). Using this idea of amphimixes, we can contour Cho's lengthy and singular sequence on anal splatter—that is, this gastroenterological anecdote—on a bodily continuum with the earlier articulatory explosions on pussies and eating babies (the icon of the endless consumption to achieve optimized futural being).

Rather than from the back of the throat, Cho speaks from the amphimixed entry points of the gastrointestinal and erotic-libidinal tracts, places that we think of as inside us but which are, from the perspective of anatomy [. . .] enclosed spaces that are nonetheless part of the outside world. [. . .] Not only this biophysiological framing but the belly techniques of Cho's concert tactilely communicate the lie of containment, the amphimixes of laughing felt in her audience's guts (as intestinal/abdominal waves of radial flexing and release, as well as pulsating respiration and vocalization coassembled). The affective style that she parasitically imitates is both advertising style and explosion/surprise; but it is also the basic biological action of peristalsis (shitting, vomiting, swallowing, passing gas).

Through critical prose informed by a feminist-inflected neuroenterological science, Wilson propositionally tells us of the agency and intelligence in the cellular materials of the wormlike intestinal tract, whereas Cho in laboratory fashion performatively stimulates that GI knowledge felt in her audience's guts, rather than having it consumed cognitively primarily through the central nervous system. Choreographing a politics pursued on the visceral or gut level, Cho's concert tactilely communicates the lie of containment, the amphimixes of knowing laughter ruminated over and sensed by the belly.

Conclusion: On Peristaltic Choreographies

In her efforts to broaden the terrain of reproductive issues to include the petrochemical industry, the science and technology scholar Michelle Murphy urges a broadened terrain of conceptualizing biological reproduction: "What counts as reproduction? Where does biological reproduction reside? . . . What is the place of industrial chemicals in reproduction?" (Murphy 2013). My argument very much accords with the spirit of Murphy's questions even as it figures US militarism (a specific armament to further petrochemical extraction and what petrochemical extraction—combustion—supports) as a key infrastructure of the denied futurism evoked in both the Thai sex worker's "pussy" and Cho's phrase "I ovulate sand." Foreclosed from the reproductive model of the heteronormative nuclear (male-headed) family, Cho's queer agency lies in her smutting up the living rooms of American households with the highly eroticized servicer of "rest and relaxation"—this Asian pussy providing gratification (gifts) free of obligations (relational ties established by reciprocal caregiving). Cho uses her humor to emphasize not only the ubiquity of biopower's disciplinary apparatus, its making her own self—rather than some authority outside herself (e.g., a sovereign father)—the very instrument of bodily regulation, but also her and other excessive, out of control, femiqueer bodies' intransigence to this clinical and empiric containment.

In terms of Cho's salience as a thinker and affective modulator of turn-of-the-twenty-first-century racialization, she does not offer any promise of justice achieved. Her concerts do not affirm that we could get to a place beyond race or beyond biopolitical contradictions. We might construe her stand-up as working on a principle that values less cognitive resolution than the dwelling in embodied contradictions. Rather than exposure leading to justice—all is right with the world as long as we expose what is wrong—Cho dwells in the inaugural moment of surprise, of antagonisms that cannot be ironed out, for example, the birth of the other entwined with the tissue morbidity of parts of the self. Reframing Asian Americanist critique in biopolitical terms, Cho's comedy instantiates a specific type of repetitive aesthetic—intensely vital *and* morbid—that formally imitates not forward progress but cyclic or helical stagings of opposite phenomena. But there are only repeated convulsive discoveries of these iterative contradictory opposites, or put another way, no subsumption and no transcendence but only peristaltic waves that push (or connect us to) the world through both ends. [. . .]

[. . .] Parallel to the cellular matter of amphimixes, Wilson argues, is the distinction between two kinds of methods or approaches, those marked by "an overriding concern with clearly demarcating causal primacy . . . as if determination is a singular, delimited event"—which she shorthands as Boolean logic, and a "more plastic model" attuned to "the amphimixed inclinations of the [biological] substrata involved" (2004, 83–84). Cho solicits her audience's amphimixed apprehension of containment's lie, even as we continue headstrong into the optimistic, anticipatory belief in endless optimization.

A Boolean version of this critical essay would conclude here with the singular assessment of Cho's message, but one wonders whether a normative message can emerge from amphimixes or a peristaltic squeezing out of both ends. Isn't the point, rather, that an amphimixed message laughs, but in revision of the medusa, not via plural heads but from the messy assemblage of bottom orifices—urethra, anus, vagina, and, yes, vomiting mouth—at the demand for a single message: the normative point or straight buildup to the end. Instead, Cho's giggling peristaltic waves intensify and wane, in waves that overlap, fold, producing whirlpools that are the wider-environmental (spatial) and durational (temporal) point that mocks the Anthropocene's sense of itself as progressing always into a more wealthy, ethical future simply because of the directionality of the arrow of historical time. The belly knows that its choreography jiggles in all sorts of directions, which may be a queer refusal of normative argumentation as political critique, but is critical nonetheless.

Notes

1 Cho's characterization of her reproductive body parts as desert-like—bereft of life—grows beyond a mere personal significance as her show progresses, staging a debate as to whether as a woman of color her reproduction is regarded as degenerate, or as an American woman her reproduction also implies an optimal to-be-protected life.

2 Amphimixis usually refers to the mode of reproduction in hermaphroditic animals, for instance, those that do not self-fertilize (i.e., reproduce by automixis) as do many hermaphroditic plants, but that mate with another hermaphroditic member of their species (an outcrossing opportunity). Elizabeth Wilson (2004), borrowing from Sándor Ferenczi (1938), uses the term *amphimixis* in a distinct other way, to refer to the shared ontogenetic cellular matter that then differentiates across embryonic development into different organs and tissues. Sites of suture or vestibular passage between distinct organ systems are sites where the prior amphimixes are remembered or more evident, so to speak. In emphasizing the similar contractile kinetics choreographing birth, digestion, evacuation, and vomiting, Cho flouts the borders between organ systems stressing what [. . .] Wilson calls their phylogenetic "amphimixis."

3 As Landecker (2011) has pointed out, biologists studying metabolism have determined that it is not simply what mothers eat while pregnant that has "epigenetic" effects on the gestating fetus but, generationally, environment (qua what and how much one eats or does not eat at what stage of the life cycle) transforms the genetic and phenotypal expression two generations down. In other words, monitoring what one consumes, while possibly intervening to optimize one's own body in the temporal duration of one's own individual life cycle, has effects on the life cycles of progeny two generations into the future. Not for nothing, then, the monitoring of consumption (what one eats) and the monitoring of pregnancy come together as particularly intense sites of political investment.

4 Aired for one season (1994–95) on ABC, "All-American Girl" was the first prime-time situation comedy to feature an all Asian-American cast.

Works Cited

Bridges, Khiara M. 2011. *Reproducing Race: An Ethnography of Pregnancy as a Site of Racialization*. Berkeley: University of California Press.

Champeau, Rachel. 2013. "Changing Gut Bacteria through Diet Affects Body Function, UCLA Study Shows." *UCLA Newsroom*, May 28, 2013. newsroom.ucla.edu/portal/ucla /changing-gut-bacteria-through-245617. aspx.

Cho, Margaret. 2001. *I'm the One that I Want*. 1st ed. New York: Ballantine Books.

———. 2005. "Title." *Margaret Cho (blog)*. www.margaretcho.com/2005/06/13/belly-dance.
Edelman, Lee. 2004. *No Future: Queer Theory and the Death Drive*. Durham: Duke University Press.
Ferenczi, Sándor. 1938. *Thalassa: A Theory of Genitality*. Translated by Henry Alden Bunker. Albany: Psychoanalytic Quarterly.
Ha, Nathan Q. 2011. "The Riddle of Sex: Biological Theories of Sexual Difference in the Early Twentieth-Century." *Journal of the History of Biology* 44 (3): 505–546.
Hird, Myra J. 2012. "Digesting Difference: Metabolism and the Question of Sexual Difference." *Configurations* 20 (3): 213–237.
Landecker, Hannah. 2011. "Food as Exposure: Nutritional Epigenetics and the New Metabolism." *BioSocieties* 6 (2): 167–194.
Lee, Rachel C. 2004. ""Where's My Parade?": Margaret Cho and the Asian American Body in Space." *TDR/The Drama Review* 48 (2): 108–132.
Machado, Lorene. 2004. Cho Revolution.
Margulis, Lynn, Dorion Sagan, and Julius Korein. 1986. *Origins of Sex: Three Billion Years of Genetic Recombination*. New Haven: Yale University Press.
Moncur, Laura. 2005. "Margaret Cho's 'F**K It' Diet." *Starling Fitness*, January 28, 2005.
Muñoz, José Esteban. 1999. *Disidentifications: Queers of Color and the Performance of Politics*. Minneapolis: University of Minnesota Press.
Murphy, Michelle. 2013. "Distributed Reproduction, Chemical Violence, and Latency." *Scholar and Feminist Online* 11.
Parks, Suzan-Lori. 1995. "Elements of Style." In *The America Play, and Other Works*, 6–18. New York: Theatre Communications Group.
Roughgarden, Joan. 2004. *Evolution's Rainbow: Diversity, Gender, and Sexuality in Nature and People*. Berkeley: University of California Press.
———. 2009. *The Genial Gene: Deconstructing Darwinian Selfishness*. Berkeley: University of California Press.
Roy, Deboleena. 2008. "Asking Different Questions: Feminist Practices for the Natural Sciences." *Hypatia* 23 (4): 134–156.
Subramaniam, Banu. 2014. *Ghost Stories for Darwin: The Science of Variation and the Politics of Diversity*. Urbana-Champagne: University of Illinois Press.
Wilson, Elizabeth A. 2004. "Gut Feminism." *Differences: A Journal of Feminist Cultural Studies* 15 (3): 66–94.

Embodiments of Safety

KANE RACE

SOME YEARS AFTER THE MEDICAL ESTABLISHMENT DECLARED HIV a chronic manageable illness, Melbourne's respectable daily treated its readers to the cartoon image of a decrepit skeleton, complete with fairy wings, guzzling pills as he staggers to the ground. The caption read "Party Animal." The illustration accompanied an article, "Dancing with Death," that accounted for what was in fact a small rise in HIV infections as follows:

> Many will not be surprised. Recent surveys in Sydney and Melbourne have shown a greater incidence of sex without condoms. Health professionals believe it is not just casual unsafe sex, but problems with people getting into relationships and having unsafe sex before both partners are tested.
>
> More to the point, however, a nexus has been found between drug use—ecstasy and speed, inextricably linked to the dance-party circuit—and unsafe sex. Young gay men are taking risks because—like other young men—they believe they are indestructible. In this case, however, gay men have had their minds altered by illicit drugs, and they assume the new protease inhibitor drug combinations will save them. [. . .] (Dow 2000)

Steve Dow treads a well-worn narrative path here. His depiction of gay men "partying on as though illicit drugs will make them forget the world outside, while prescription drugs will save them from the threat of the

---

Excerpted from Kane Race, "Embodiments of Safety," in *Pleasure Consuming Medicine: The Queer Politics of Drugs*, pp. 137–163; 214–219.

virus" is the very image of technological and consumptive excess. Linked to the Internet, to drugs, and to a commercial scene, gay men are typified as self-destructive narcissists, technologically fixated, partying themselves to death. Dow even prefaces his account by alluding to its cultural availability: "Many will not be surprised." Surprise is indeed misplaced, but not for the reasons the author would have us confirm. The extraordinary levels of behavior change sustained in the first decades of the epidemic were adopted in response to a far more threatening prognosis. Gay men are no longer quite "dancing with death," but "with chronic manageable illness" (an epigraph that admittedly lacks punch). I don't mean here to recommend the acquisition of chronic illness, simply to observe that, given these changed conditions, some revaluation of risk is comprehensible—to be expected, even. But rather than give this point any serious consideration, commentators around the world preferred to depict gay men as self-consumed hedonists who had forgotten the moral lessons of AIDS. [. . .]

[. . .] As HIV enters a new, more manageable phase, illicit drugs appear to have taken up the slack in the public narrative of just desserts that has come to haunt gay life. [. . .] What's more, the moral occasion tends to take precedence over effective communication of the realities of HIV risk. Thus while Dow makes reference to the fact that many (in Australia, almost half of) new infections occur as an effect of placing too much faith in the ostensible security of conjugal intimacy, note how quickly this fact is subordinated to the thrust of the primary narrative. ("More to the point . . . a nexus has been found between drug use . . . and unsafe sex." What "point" makes this nexus more significant exactly?) [. . .]

Illicit drugs have in fact always been part of the environment in which both risk and safety (with respect to HIV transmission) have been practiced (see for example, Lewis and Ross 1995). Since the introduction of HAART, however, they have increasingly been cast in this more determinative role. This delivers both HIV and drug education to the custodians of public morality—in either the mode of moral condemnation (declaiming "abuse") or else corrective intervention (treating "addiction"). By positioning drugs as causative of HIV risk, these accounts present HIV as an outcome of personal acts of moral transgression. But HIV transmission has very little to do with the state of personalities and a lot to do with particular activities and relations between bodies. In this chapter I will offer an alternative account of what has worked in HIV prevention and drug education. My argument is that the achievements of these fields suggest the utility of a distinction between embodied ethics and normative morality—a distinction that has allowed health workers to articulate and engage the embodied pleasures of endangered groups, rather than deny them. Of course it makes a certain amount of intuitive sense that health interventions are most effective when phrased in terms of the values, media, and sources of authority that are respected by those they seek to address. But what complicates this task when the risk involves "illicit" activities (such as homosexuality or drug use) is that it is precisely the provision of this sort of material that is precluded by public morality.

Ethics and Technique

Though they have developed into distinct fields of practice and policy, both gay men's HIV education and harm reduction had as their initial conditions of possibility the highly moralized climate of the AIDS crisis. In these circumstances, these fields strategically adopted a "value-neutral" stance in the field of public policy. This involved a rhetorical distinction between *scientific facts* and *moral values*. When moral reactionaries condemned homosexuality, activists could appeal to the scientific fact that HIV could be prevented by the use of condoms. When conservatives objected to the institution of needle exchanges, practitioners could appeal to the public health objective of HIV prevention. Of course, these "rational" responses were also "moral" responses in fact. They claimed a moral entitlement to health for homosexuals, sex workers, drug users, and other marginalized subjects. [. . .]

At the same time, there is a sense that the failure to articulate a more open stance on pleasure leads to shortcomings in the position of harm reduction. [. . .] Pat O'Malley and Mariana Valverde point to the absence of pleasure as an explanation for drug use in harm-reduction education materials to argue that the "more or less explicit model of the subject deployed in harm minimization is that of the rational choice actor who will perform the felicity calculus" (2004, 36). Harm reduction is situated within a highly technical and rationalist framework that proposes the objective calculation of risks and harms, producing, for example, an "enormous amount of attention [to] the mechanics of drug administration"

(O'Malley 1999, 200). However, I suggest that the technical aspects of safe practice are important precisely because they engage the bodily *hexis* of those at risk. In this sense, they are not neutral or divorced from the field of value, but culturally embodied and experimentally cultivated. [. . .] Now, one of the crucial innovations of both HIV education and harm reduction alike has been that, rather than holding out for disembodied ideas of the good, they have met people at the level of concrete embodied practice. This can be as simple as distributing technical devices such as condoms or clean syringes at critical moments so that people can do more safely what they were going to do anyway. Or it can take the form of a reflexive discussion of bodily situation and technique. The crucial thing at any rate is not the blanket reinforcement of abstract norms, but how certain techniques have been incorporated into bodily practices, situations, and styles. [. . .]

[In the second and third volumes of *The History of Sexuality*, Michel] Foucault conceives the possibility [. . .] of what Jane Bennett calls an "experimentally *cultivated* responsiveness" (1996, 654). If his earlier work demonstrates the pervasiveness of socially imposed discipline, this later work suggests a slim margin of possibility between discipline imposed and practices of self-elaboration, yielding what Foucault calls "technologies of the self" (655). [. . .] In his subsequent work on "care of the self," he wants to conceive forms of care and self-relation that could pry themselves away from normative determinations where necessary, but retain some form of ethical stylization (Foucault 1990, 2012).

There are echoes of this later work in Douglas Crimp's argument, made in the early days of the AIDS epidemic: "We were able to invent safe sex because we have always known that sex is not, in an epidemic or not, limited to penetrative sex. Our promiscuity taught us many things, not only about the pleasures of sex, but about the great multiplicity of those pleasures. It is that psychic preparation, that experimentation, that conscious work on our own sexualities that has allowed many of us to change our sexual behaviours—something that brutal 'behavioural therapies' tried unsuccessfully for over a century to force us to do—very quickly and very dramatically" (1987, 253). For Crimp, safe sex is an outgrowth of embodied improvisation. The initial appearance of AIDS among gay men and drug users in the West gave rise to dogmatic calls for abstinence, monogamy, and quarantining.[1] These calls did not just come from moral reactionaries; they were the logical extension of prevailing mechanisms of public health and medical science (Patton 1990; Kippax and Race 2003; Ballard 1998). They were premised, that is, on rationalities that positioned those at risk as objects of government, rather than subjects of their own care. Not only did endangered groups withstand these calls, but they also devised strategies that successfully prevented transmission without demanding a "return" to normative forms of sexual and corporeal life. [. . .]

The findings of social science certainly bear out a relation between embodied subjectivity and the strategies that have been successful with respect to HIV prevention. While gay men largely ignored the moral calls for abstinence and monogamy, between 1986 and 1996 there was an uptake of prevention strategies grounded in the embodied styles of those at risk, including the use of condoms for anal sex, an expansion of the sexual repertoire in terms of the adoption of relatively safe sexual practices, and, when HIV testing became common, negotiated safety (where regular partners of the same HIV status negotiate an agreement with respect to sex that happens inside and outside the relationship) (Kippax and Race 2003). Not only did gay men and drug users choose some strategies and eschew others, but there is also evidence for the effectiveness of these strategies. As Susan Kippax and I have shown (2003), the strategies that have been most successful have been those based on a mutually acceptable description of safety from the perspectives of official science and the embodied positions of those at risk. "Gay men, injecting drug users, and, to some degree, heterosexuals moved to make their practices safe—by modifying and building on them, not by abstaining from or eliminating them" (6). The picture of agency that emerges from these data is neither that of the self-knowing, decisional subject whose actions are the result of rational choice and disembodied control, nor the behavioral automaton that merely reproduces given norms, but rather something much more akin to embodied ethics. Safe practice is an outcome of embodied habits, cultural memory, and sedimented history; but it also depends, to some extent, on ethical improvisation and modification. [. . .]

Uses of Medicine

One of the advantages of recognizing embodied agency in the field of HIV has been how it elucidates the effect of medical technologies on prevention. When a decline in consistent condom use became apparent among gay men in the 1990s, many epidemiologists spoke of recidivism or non-compliance. Research that was more sensitive to the social contexts of gay sex was able to detect that gay men were using the medical technology of HIV testing to enable relatively safe, unprotected sex within regular relationships (Kippax et al. 1993). A number of gay men were discarding condoms for sex within seroconcordant regular relationships, but continued to use them within casual contexts. Australian educators responded to this finding by conducting specific education in order to make this strategy safer, termed "negotiated safety" (Kinder 1996). Since it is unreasonable to expect adult individuals not to inform and avail themselves of medical knowledge, the most feasible option open to educators seemed to be to make the existence of this practice known to affected individuals and enhance its safety. [. . .]

The introduction of HAART saw further changes to the sexual cultures of gay men—in particular rises in unprotected sex (though these rises were not necessarily accompanied immediately by increases in HIV infections) (Ven, Rawstorne, et al. 2002). Qualitative research conducted in 2000 identified the existence of a range of considered strategies other than condom use and negotiated safety in gay men's accounts of unprotected sex (Rosengarten, Race, and Kippax 2000; Kippax and Race 2003; Race 2003). These included the use of viral load test results to estimate the risk of infection, withdrawal before ejaculation, disclosure of HIV status, and the adoption of an insertive or receptive position when engaging in unprotected anal intercourse, depending on HIV status. Subsequent quantitative analyses confirmed the use of these strategies to reduce risk among Sydney gay men (Ven, Kippax, et al. 2002). When engaging in unprotected anal intercourse, HIV-negative men in casual encounters were more likely than not to adopt the insertive role, while HIV-positive men were more likely to be receptive. [. . .]

Whatever the relative efficacy of these strategies, their existence contradicts the stereotype of the gay cultural dope. [. . .] What seems to be the case is that the introduction of HAART provided the conditions of possibility for a limited process of revaluation of risk among gay men. It remains the case that a majority of gay men report using a condom in every instance of anal sex in a defined interval, but an expansion of relatively well-informed prevention ethics that do not equate with the condom ethic is also apparent (Ven, Rawstorne, et al. 2002). This is not a picture of people throwing caution to the wind. It is a picture of gay men appropriating medical knowledge to craft a range of considered strategies that attempt to balance sex and the avoidance of risk. These strategies are not foolproof. But many of them are scientifically plausible on the basis of epidemiological evidence, indicating a relatively informed engagement with scientific and probabilistic reasoning in gay men's sexual accounts and practice. [. . .] These strategies adapt medicine to fit the cultural desires of subjects for pleasure and safety. What emerges is a field of complex evaluation that does not always rank long-term health over everyday desires for intimacy, sensuality, and pleasure.

The use of medicine within gay sexual repertoires raises a number of challenges for HIV educators, not least the challenge of enhancing the safety of these practices. This entails more than supplying people with scientific information about the risks of various practices and more than a celebration of agency. Upon the identification of the workings of medicine in gay sex, the field of HIV prevention in Australia became prolific with accounts of "sophisticated gay men making complex decisions to reduce risk." At times, these descriptions seemed to verge on a populist romanticism that celebrated every instance of gay men's appropriation of medicine as though it were inherently safe. [. . .] It seems to me that the reason for attending to embodied agency in the field of health promotion is not simply to celebrate the endlessly inventive practices of everyday life or sex (though well we might). It is to provide much-needed information about the cultural conditions in which particular dangers and possibilities—both social and physical—take shape. Some of these possibilities arise from the very practices of differentiation that are adopted to promote safety (such as those that discriminate between HIV-positive or HIV-negative sexual partners, or, as I will discuss later, between different forms of drug use). Health promoters need to monitor the cultural effects of these practices and discourses carefully and, when their effects become problematic, respond to them.

A Taste for Drugs

[. . .] Drug choices take place within a whole world of meaning and cultural value. [. . .] Choices of chemical modification entail processes of cultural classification and discrimination. They are ingrained in the body as habit, such that they feel deeply visceral and compelling. But they link also to sentient mythologies of taste, performance, affect, mood, and sociability.

It is also the case that a concern for the body—its safety and its limits—can play a part in these processes of cultural preference and distinction. Robert Reynolds gives a sense of this in his account of the activities leading up to the Mardi Gras party: "I love this part of the evening. . . . You gather with friends, divide up the drugs, compare booties and pill dropping schedules, and ease into the night with prickles of anticipation. . . . It's hard not to feel the frisson of risk, deliciously illicit yet carefully managed" (2002, 70). Here care is part of a pleasurable social activity involving habitual acts of planning and comparison that become second nature to participants. In their ethnographic study of drug use among Sydney gay men, Erica Southgate and Max Hopwood document the existence of what they call a "folk pharmacology" that informs safe drug practice in this setting. This is "manifested in judgments, decisions, and practices" that define and delineate what types of illicit drug use are considered desirable and acceptable, and what not, and includes considerations of drug effects, routes of administration, and the risks and harms of different forms of use (2001, 325). Notions of competent drug use were embedded within the "mastery of a multitude of practices that make up 'partying'"—practices that include "choosing and wearing the right clothes; developing and displaying your body in a particular manner ('attitude'); dancing according to a certain style; 'cruising' for sex using a variety of understood bodily codes; and taking specific types of drugs and combinations of these drugs according to a number of folk rules designed to minimize the risk of one becoming a 'messy queen'" (325). Being messy meant being "out of control" on drugs, and the authors list a number of practices and procedures intended to prevent this possibility. These include considerations of drug selection, timing, and environment, as well as measures taken around the anticipation of particular physical or emotional states. "Becoming messy meant ruining the pleasure of a good night out not only for oneself, but for friends and lovers. It also meant the potential forfeiting of a rather substantial investment made in terms of the purchase of costume, tickets, and drugs" (325). In other words, considerations of care and safety are often inextricable from the field of value that surrounds the drug experience. The authors identify a hierarchy of value operating within this field involving distinctions between, for example, competent versus "messy" use, different routes of administration, and of course "good drugs" (in this sample ecstasy, speed, GHB) and "bad drugs" (heroin, alcohol). Obviously, such distinctions and forms of cultural discrimination are not without their problems. For example, the authors describe how value judgments surrounding some modes of use (injecting) served to dissuade participants from that activity, but also had the potential to socially isolate users of that mode. We can nonetheless see that drugs are linked to specific practices of pleasure and forms of sociability. They are used to produce particular contexts of interaction, pleasure, and activity. This entails a host of tiny routines and decisions about method, context, disposition, and desire. And these decisions are linked to a series of moral judgments around desirable sensations, uses, outcomes, and behavior. [. . .] Most importantly, we can see that specific practices of care and attention are being brought to the question of how to use drugs, such that considerations of safety appear as part of a concern to *maximize* pleasure, rather than standing in direct opposition to it.

It is precisely this sense of care as a potentiality in the body that many HIV and harm-reduction initiatives have sought to actualize. Recognizing the situated nature of the most serious harms, they aim to identify and enhance the processes through which people look after themselves. This creates an emphasis on technique, proposing a differentiation in the field of "uses" that is neatly summed up by the saying "how to do more safely what you were going to do anyway." In practical terms, this has generally involved acting on the environments and situations in which specific risks are taken—for example, the provision of clean syringes, the creation of injecting rooms, or the regulation of sex venues so that they provide condoms and lubricant. But when it comes to a more explicit articulation of pleasure, these initiatives tend to be less forthcoming. Thus it is with some surprise that Pat O'Malley and Mariana Valverde note the absence of any discussion of pleasure in harm-reduction educational materials (2004, 36). [. . .] If drug consumption is analogous to consumption in general, its intelligent use would seem to necessitate more opportunities to consider and evaluate the qualities of pleasures, not less. By evacuating pleasure from the field of concern, these materials risk reproducing a sense of the body as a freely chosen fate.

Sensitive Material

HIV-prevention education that has sought to engage the characteristic pleasures of those at risk tends to work in a register that is at once technical and ethical. It is technical because it concerns itself with "how to" questions. And it is ethical because it recognizes that these questions are always embedded within a practical horizon of concerns that extends beyond a simple concern around infection to take on questions of "how to be in an environment and in relation to other people" (McInnes, Race, and Bollen 2002, 31). Two recent resources from HIV organizations in Australia provide good examples of this style of education: *When You're Hot You're Hot* and *HIV + Gay Sex*. Both of these booklets address themselves to particular target audiences—potential users of sex-on-premises venues and HIV-positive gay men, respectively. They are distributed through select venues, and their content takes the form of a series of "how-to" tips: how to use sex venues and how to be sexual as an HIV-positive gay man. This is a familiar enough genre in popular culture—the stuff of infotainment—though the topics these resources deal with are far less likely to be so generally available.[2] Already we are beyond the comfort zone of a position that would insist that people generally should not have casual sex, and that HIV-positive people should not have sex at all. These materials are noteworthy because they highlight the fact that ethical codes and principles come into play even in situations that are constituted as "beyond the bounds" of normative morality. They evaluate well; this is not surprising, since they answer questions that people have been made afraid to ask in public. Both resources contain advice on condom use, risks of transmission, and so on. But they situate this advice within the world of concerns as it appears to these specifically situated subjects. So in *HIV + Gay Sex*, we don't simply have "how to use a condom." We have information on how to maintain a fulfilling sex life, how to disclose your status in a way that feels safe, how to feel sexy when you're constituted as infectious, how to negotiate a relationship with a negative lover, and so on. In *When You're Hot*, we have how to find your way around a venue, how to cruise for sex, how to communicate your desires and limits without talking, how to socialize in this space, and so on. This is very practical, very technical advice. In a sense it is no different to handing out a condom at a dance party. But it is conducted in a more textured or readerly medium. [. . .]

These resources work by making what are perhaps, for some, *unthinkable* domains of activity available to thought and practical consideration. They can be understood as technologies of government—as well as of the self—in that they aim to make particular fields of experience thinkable in certain ways (see, generally, Rose 1998). It would be naive to claim these materials operate romantically outside the scope of power. [. . .] At the same time, it seems necessary to insist that the sorts of subjects and choices that these resources support are not subjects or choices that normative morality can tolerate easily or even allow to exist (which may explain the restrictive conditions within which these resources circulate). These initiatives can be characterized as engaging their audiences at the level of embodied ethics (as distinct from morals). They direct themselves to particular scenes in which people are making themselves into subjects. Note that the moral principle around nontransmission of HIV is still there. But it has been converted into technical advice that knows and affirms the embodied worlds of the audience each resource anticipates. [. . .][3]

Conclusion

The care practices and corporeal pedagogies outlined in this chapter exist in a tense relation with hegemonic prescriptions around corporeal practice, modes of consumption, and relations to medicine. This tension circumscribes and constrains the production and circulation of educational materials that are grounded in the embodied practices of endangered groups. As well as obstructing effective education, one effect of this tension is to sensationalize social deviance as risk. Practices that may actually be safe with respect to HIV transmission, for example—or which may emerge precisely as attempts to find workable ways of avoiding HIV transmission, or construct collective contexts for the elaboration of practical ethics grounded in existing embodied practice—become sensationalized on the basis of the deviation they represent from corporeal norms and materialize as thrilling instances of transgression. [. . .] Apart from demonizing individuals, the danger of this exercise of power is twofold. First, it eroticizes transgression for its own sake. Second, it promotes both public and personal misrecognition of the possibilities of care that actually inhere in given bodily practices. [. . .] Normative constructions of responsibility work here to undercut the practical or embodied ethics that are being elaborated in these contexts, with the effect of spectacularizing them only as risk. [. . .]

[. . .] To interpret the sex and drug practices considered [here] as only about transgression and escape is an inattentive reading of the pleasures and possibilities of subordinated bodies. Certainly, when it comes to the well-being of such bodies, it is often necessary to transgress social prescriptions around what it is possible to say and do. These social prescriptions are loaded against the adoption of more careful and attentive postures toward the embodied experience of subjects thereby deemed illicit, as I have argued throughout. The tension is particularly acutely felt in the fields of HIV prevention and drug harm reduction today, where normalizing therapeutic discourses are consistently cited to block the development of practical sex and drug pedagogies that admit of, and seek to work with, pleasure. Indeed, the elaboration of "care of the self" and "bodies and pleasures" within critical health practice paradoxically involves a struggle against individualization, particularly the forms of privatization associated with the neoliberal state. The social making of the "bad example" effectively sensationalizes social deviance as willful risk (or else compulsive pathology), in a move that produces unendurable blind spots, enforces blockages, and preempts more grounded possibilities of care. Given such circumstances, perhaps the most viable way to counter the unhealthy individualization effected by exemplary power is to move—as impersonally as possible—through it.

Notes

1 From an article by William F. Buckley Jr. printed in the *New York Times* in 1986: "Everyone detected with AIDS should be tattooed in the upper forearm, to protect common-needle users, and on the buttocks, to prevent the victimization of other homosexuals." Quoted in (Watney 1997, 44)

2 I'm indebted here to Gay Hawkins's argument in "The Ethics of Television" (2001).

3 These particular resources are the product of qualitative inquiries conducted by educators and university-based cultural researchers trained in contemporary theories of sexuality and the body.

Works Cited

Ballard, John. 1998. "The Constituting of AIDS in Australia: Taking 'Government at a Distance' Seriously." In *Governing Australia: Studies in Contemporary Rationalities of Government*, edited by Mitchell Dean and Barry Hindess, 125–138. Melbourne: Cambridge University Press.

Bennett, Jane. 1996. "'How is it, Then, That We Still Remain Barbarians?' Foucault, Schiller, and the Aestheticization of Ethics." *Political Theory* 24: 653–672. doi: 10.1177/0090591796024004003.

Crimp, Douglas. 1987. "How to Have Promiscuity in an Epidemic." *October* 43: 237–271. doi: 10.2307/3397576.

Dow, Steve. 2000. "Dancing with Death." *Age*, 4 October 2000.

Foucault, Michel. 1990. *The Care of the Self: The History of Sexuality* 3. Translated by Robert Hurley. London: Penguin.

———. 2012. *The History of Sexuality, Vol. 2: The Use of Pleasure*: Knopf Doubleday Publishing Group.

Hawkins, Gay. 2001. "The Ethics of Television." *International Journal of Cultural Studies* 4: 412–426. doi: 10.1177/136787790100400403.

Kinder, P. 1996. "A New Prevention Education Strategy for Gay Men: Responding to the Impact of AIDS on Gay Men's Lives." XI International AIDS Conference, 1996.

Kippax, Susan, June Crawford, Mark Davis, Pam Rodden, and Gary Dowsett. 1993. "Sustaining Safe Sex: A Longitudinal Study of a Sample of Homosexual Men." *AIDS* 7: 257–263. doi: 10.1097/00002030-199302000-00015.

Kippax, Susan, and Kane Race. 2003. "Sustaining Safe Practice: Twenty Years On." *Social Science & Medicine* 57: 1–12. doi: 10.1016/S0277-9536(02)00303-9.

Lewis, Lynette A., and Michael W. Ross. 1995. *A Select Body: The Gay Dance Party Subculture and the HIV/AIDS Pandemic*: Burns & Oates.

McInnes, David, Kane Race, and Jonathan James Bollen. 2002. *Sexual Learning and Adventurous Sex*. Sydney, Australia: University of Western Sydney Press.

O'Malley, Pat. 1999. "Consuming Risks: Harm Minimization and the Government of 'Drug-Users.'" In *Governable Places: Readings on Governmentality and Crime Control*, edited by Russell Smandych. Aldershot, U.K.: Dartmouth.

O'Malley, Pat, and Mariana Valverde. 2004. "Pleasure, Freedom and Drugs: The Uses of 'Pleasure' in Liberal Governance of Drug and Alcohol Consumption." *Sociology* 38: 25–42. doi: 10.1177/0038038504039359.

Patton, Cindy. 1990. *Inventing AIDS*. New York: Routledge.

Race, Kane. 2003. "Revaluation of Risk Among Gay Men." *AIDS Education and Prevention* 15: 369–381.

Reynolds, Robert. 2002. "Through the Night." *Meanjin* 61: 67–73.

Rose, Nikolas. 1998. *Inventing Our Selves: Psychology, Power, and Personhood*: Cambridge University Press.

Rosengarten, Marsha, K. Race, and S. Kippax. 2000. "'Touch Wood, Everything Will Be Ok': Gay Men's Understandings of Clinical Markers in Sexual Practice." Sidney: National Centre in HIV Social Research.

Southgate, Erica, and Max Hopwood. 2001. "The Role of Folk Pharmacology and Lay Experts in Harm Reduction: Sydney Gay Drug Using Networks." *International Journal of Drug Policy* 12: 321–335. doi: 10.1016/S0955-3959(01)00096-2.

Ven, P. Van De, S. Kippax, J. Crawford, P. Rawstorne, G. Prestage, A. Grulich, and D. Murphy. 2002. "In a Minority of Gay Men, Sexual Risk Practice Indicates Strategic Positioning for Perceived Risk Reduction Rather than Unbridled Sex." *AIDS Care* 14: 471–480. doi: 10.1080/09540120208629666.

Ven, P. Van de, P. Rawstorne, J. Crawford, and S. Kippax. 2002. "Increasing Proportions of Australian Gay and Homosexually Active Men Engage in Unprotected Anal Intercourse with Regular and with Casual Partners." *AIDS Care* 14: 335–341. doi: 10.1080/09540120220123711.

Watney, Simon. 1997. *Policing Desire: Pornography, AIDS and the Media*: A&C Black.

Consent, Capacity, and the Non-Narrative

AMBER MUSSER

IN OCTOBER 2014, THE STATE OF IOWA CHARGED HENRY RAYHONS with third degree sexual assault for having intercourse with his wife Donna, who was deemed unable to consent to sexual activity due to her advanced state of dementia. Rayhons was later acquitted, but the case set off an intense debate about dementia and intimacy (Belluck 2015; Brenoff 2015; Singai 2015).[1] Alongside this case, I position *John Doe v. Occidental College*, which John Doe (a pseudonym) filed in February 2014 for his expulsion following a charge of sexual misconduct against him for an incident in which both he and Jane Doe (another pseudonym) were incapacitated due to alcohol and could not remember having sex, though both appeared to consent in the moment (Hess 2015; Dorment 2015).[2] Through a reading of these cases, I argue that contemporary understandings of consent require a nonpathological subject, as consent is considered impossible for those deemed "incoherent" or "incapacitated." In both cases, notions of pathology were in play, although, on the one hand, we have dementia's creeping and irreversible diseased forgetting, which is tethered to the idea of a permanent pathological state, while, on the other hand, the episodic memory loss that alcohol- or drug-induced intoxication produces speaks to an understanding of pathology that is temporary and situational. The reliance of notions of consent on a nonpathological, coherent subject with capacity effectively renders consensual sexuality impossible for certain nonnormative subjects, both those understood as permanently pathological and those understood as situationally pathological. To decouple discussions of consent from notions of pathology, coherent linear selfhood, and capacity, I turn to an unlikely source, the journals of Gary Fisher, to revalue aberrations of memory and the incoherence of selfhood. Fisher's journal entries provide an example of immediate desires, and I argue that in the immediate we can register Donna's desire (and the desire of others who have been deemed pathological and thereby nonautonomous) on its own terms as something available and nonpernicious, regardless of its narrative incoherence. In the immediate, we can also register the sensuality and desire at work behind moments when the self is incoherent (or the situationally pathological, as in the case of John and Jane Doe).

Consent has become the dominant framework for understanding various forms of trespass. In his theorization of consent, Joseph Fischel argues that it can function as a way to ascertain what constitutes a violation. In describing its importance in sex law, Fischel writes, "As a prominent metric dividing permissible sex from impermissible sex, or sex from rape, nonconsent is normatively superior to considerations of marital status, age (without qualifiers), gender, or race in historically oppressive examples. Consent or its absence imperfectly but importantly tracks the sexual autonomy rights of individuals" (2013, 57). What Fischel's discussion of the use of consent makes clear is the way that consent not only aligns parties in agreement about what should take place (which is to say that there is a linear progression of behavior); consent also imagines that both parties are *able* to agree. That is to say, all parties are autonomous and agential. This valuation of autonomy and agency has been central to our current sexual landscape, as evidenced by the feminist debates about whether participation in sadomasochism or pornography signaled false consciousness or sexual liberation. Though I have no interest in rehashing these debates, I mention them here to give the context that led to consent's emergence as a central value in theorizing sexuality. Consent offers the promise of sex without harm and the possibility of embracing what Gayle Rubin calls "benign sexual variation," as a step toward the "legal and social legitimation of consenting sexual behavior" (2002, 167). Within this framework, the liberal subject is the exemplary sexual actor, and actions that occur outside of this understanding are part of the territory of abuse. Further, consent implies a contract of sorts between parties based on certain assumptions about narrative, linear progression, and normativity. It is the framework, fantastical though it may be, for understanding how different people can occupy the same space and be together. Importantly, the embrace of consent privileges an ethics of nonviolation at the core of interactions with others, and it puts forth the liberal subject as the actor capable of saying no. Even affirmative consent laws, which have been introduced in order to move away from a model of explicit prohibition, because they are assumed to put undue burden on the victim of sexual assault, are embedded in this framework of liberal subjectivity. In statute SB-967, passed in September 2014, the state of California required all colleges and universities to adopt an affirmative consent model. They defined affirmative consent as follows:

> "Affirmative consent" means affirmative, conscious, and voluntary agreement to engage in sexual activity. It is the responsibility of each person involved in the sexual activity to ensure that he or she has the affirmative consent of the other or others to engage in the sexual activity. Lack of protest or resistance does not mean consent, nor does silence mean consent. Affirmative consent must be ongoing throughout a sexual activity and can be revoked at any time. (California Senate, 2014)

In moving toward prioritizing in-the-moment, conscious, voluntary agreement to proceed with sexual activity, advocates of the bill also wanted to move away from an ideology that sex is "taken" from someone else and toward an ethos of mutual agreement through a variety of means, including the nonverbal. However, in practice, this still places an emphasis on the production of a coherent self and thus renders consent impossible for those deemed "incoherent" or "incapacitated." Consent is negated if any form of pathology is presented.

Embedded in the legal and ethical quagmire over these issues of consent is the assumption that only those who can present a linear narrative of the self and who act in accordance with that narrative are capable of consent. At stake in these cases is not only what constitutes consent, but the assumption of a normative form of selfhood. When selfhood becomes incoherent or pathological and the framework of consent becomes murky—in these cases, because of dementia or alcohol and drug use—the possibility for consent is denied, producing only pathologized victims and criminalized offenders.

Rayhons, a former Iowa state legislator, was charged (and acquitted) with sexually assaulting his wife Donna, who was living in a nursing home with severe dementia. Henry and Donna—each previously widowed—married in 2007 and enjoyed a happy marriage even as Donna's health deteriorated due to early-onset Alzheimer's. After incidents in which Donna was found wearing only lingerie under a coat far from her home, her family decided to place her in a nursing home. Henry visited often and the two maintained a sexual relationship. However, once Donna became unable to recall the words for sock or blue, staff members told Henry that Donna was no longer able to consent to sex. Eight days after that incident, Donna's roommate reported hearing sexual sounds from her room and saw Henry deposit her underwear in the laundry bin. Alerted to the possibility that Donna and Henry had engaged in intercourse, her children banned Henry from visiting their mother and had him arrested for sexual assault. Though he initially admitted engaging in sexual activity, during his trial Henry argued that he had confessed under duress and that during the day in question intercourse did not, in fact, occur; instead, he had rearranged her position on the bed to make her more comfortable. While he was later acquitted, his trial has become illustrative of the ethical quandaries that attend to consent and dementia.

In the Rayhons case, several additional layers complicate the relationship between selfhood and consent. First, the fact that Rayhons was charged with sexual assault within the context of marriage should be read as the result of feminist legal victories regarding marital rape, in which evidence of prior consent—marriage—is not evidence of current consent, which resulted in the criminalization of marital rape in all fifty states by 1993 (Hasday 2000). Following from this, Donna's prior consent to sexual activity cannot be used as evidence toward her possible consent in the current moment. What complicates this severing of the past and the present is that Donna had been deemed incapable of exercising autonomy because of her Alzheimer's. When the staff at the nursing home argued that Donna could not consent because she was unable to access much of her working memory—she could not access language, nor could she remember how to properly use the sink or other facilities—it was because her actions lacked meaning within a larger frame of signification. By arguing that Donna could no longer consent to sexual activity, the staff at the nursing home were putting forth a theory that tethered consent to a version of selfhood that was stubbornly attached to the liberal individual and the ideal of a coherent narrative of self. Because dementia had made

Donna's access to a linear narrative of self through memory impossible, consent was always outside of her purview. The illegibility of Donna's current desires was further reinforced when a judge appointed temporary guardianship to her daughter, Suzan, after the incident (Yost 2014). As a proxy for Donna, Suzan's choices, however, were not guided by Donna's immediate needs or desires, but by a commitment to authenticity, which "refers to the congruence between a person's values (including beliefs, commitments and relationships) and a decision" (Kim 2011, 410; Brudney 2009; Sulmasy and Snyder 2010). This form of representation meant that Donna's linear long-term memory was assumed to be more in keeping with her desire and vision of self than her present actions.

From the framework of consent, Donna's present is pathological in that she cannot suture it to her past. Her in-the-moment decisions (even as they may be voluntary and conscious) cannot be considered agential because she cannot make sense of them within a longer narrative of self. This effectively precludes Donna from being able to engage in sexual activity without criminalizing her partner. Her inability to consent and the surveillance that is part of her care for dementia radically inhibit her ability to express her sexual desires. Some bioethicists argue that this unduly punishes people with dementia, who still desire physical contact even if consent cannot be verbalized and a person's other capabilities are much altered. In her commentary on the case, Tia Powell, director of the Montefiore Einstein Center for Bioethics at the Albert Einstein College of Medicine, told *New York* magazine, "There's nothing about being cognitively impaired that means you wouldn't necessarily appreciate being connected with other people through both nonsexual and sexual means" (Singai 2015). For Powell, this means allowing and monitoring sexual activity to ensure that it doesn't cross the threshold into abuse and that it "seems to bring comfort or enjoyment to the individual with dementia, and . . . isn't causing significant harm to others" (Singai 2015). While Donna's long-term memory and immediate actions might suggest that she might still desire sexual activity with her husband, her dementia has introduced a schism between that self and her current self. Donna's positioning as pathological renders her forever unable to consent.

In their theorization of consent and disability, Joseph Fischel and Hilary O'Connell argue for a redefinition of sexual autonomy, so that it is considered a human capability rather than a choice. They emphasize the importance of thinking about sexual acts relationally and use the framework of capability to think beyond individuals and toward "institutional reforms, statutory provisions, and alternative divisions of labor and play that cultivate human (sexual) flourishing" (2016, 468). Their framework allows for individuals who have been deemed incapacitated to engage in sexual activities without criminalizing their partners. In addition to suggesting reforms to increase access to sexual information and opportunities, education about sexuality, and provisions for sexual assistance, they argue that their framework of sexual autonomy—"as dynamic, aspirational, achieved, and relational rather than binary and keyed to a rationality threshold—readmits certain subjects (like, say, some teenagers and persons with disabilities) historically ejected from its ambit" (2016, 524). This move away from consent toward a differently defined sexual autonomy opens doors for thinking about sexuality vis-à-vis people who have been deemed permanently incapacitated. It does not, however, change things for those who are situationally incapacitated, as discussed below.

In the case of John Doe consent is also at issue, but the discourse of pathology presents very differently. Here, the aberrations in memory are about an episode of forgetting, which is the potentially undesired result of choosing to become intoxicated, rather than dementia's more progressive and cumulative lapses in memory. Because the relationship to self is not permanently pathological, pathologization becomes temporary and situational. The event that caused John Doe to sue Occidental College stems from an incident from fall 2013 when two first-year students, John and Jane Doe, engaged in sexual intercourse while both were incapacitated by alcohol. By the time John and Jane met at an impromptu dance party, both had been drinking for hours. John was drunk after a water polo initiation party, and Jane had been drinking at a small party with friends. Jane initiated contact with John; he reciprocated; her friends brought her back to her room, but after they put her to bed, she returned to John's room to have sex. The next morning neither party could remember the event, though John found a used condom and Jane's earrings in his room. They met up and reconstructed the evening by looking at their text messages and speaking with friends. Once they became aware of what had transpired, Jane became uncomfortable with the series of events and initiated a sexual assault complaint. Occidental agreed that Jane could not have consented to sexual activity and held John responsible, expelling him for violating its sexual misconduct policy. John, who had framed the encounter as a drunken mistake, sued Occidental for gender discrimination because his state of drunkenness and incapacitation also made it impossible for him to consent to the sexual encounter. The case, whose outcome is still pending, set off an avalanche of controversy around what happens in cases when neither party could legally consent to sex, around gender and sexual appetite, and around the relationship between self and memory.

There are clear gender norms at operation here—Jane is positioned (and positions herself) as situationally pathological, while John is positioned as criminal. While Donna's understanding of herself is understood to be forever out of reach, once John and Jane have sobered up, they can gauge their encounter against an understanding of their long-held desires. Though both parties were technically unable to consent to sex due to their levels of inebriation, their in-the-moment consent was invalidated when it was positioned against Jane's long-term memory. In her reporting on the case, Amanda Hess writes that Jane took the case to Occidental because it felt like an aberration of her normal behavior. Hess writes, "Because Jane blacked out on the night of September 7, 2013, she doesn't remember how she felt as the events unfolded—just how she felt before and after. 'The thing is I have no clue what I was thinking,' Jane later told investigators. 'I would have never done that if I had been sober. . . . I don't know what was going through my head.' . . . Jane's friends and an Occidental advisor counseled her that her experience was consistent with rape. So she filed a complaint" (2015). Since neither Jane nor John remembers the encounter or consenting to it, Jane's feelings of regret and trauma, which stem from her understanding of herself in relation to her narrative of self, is prioritized while spontaneous actions that are not indexed to this narrative are pathologized as aberrant.

At the same time, we see the importance of futurity in producing Jane as situationally pathological. In Donna's case, Alzheimer's is seen to foreclose futurity because the pathological state is permanent, but in Jane's case, the future comes into play. We see this in Jane's dismay that her first experience of sexual intercourse unfolded as it did. According to Hess, she told investigators that "as a 'hopeless romantic,' she had hoped her first time would be different" (2015). What we see in this statement is not only regret, but a tethering of disappointment to a future that did not happen—a "first time" that would be bound up in a "happily ever after"—a capacity that was not enacted due to an evening of incapacity. This retrospective assessment of her evening further illuminates the gendered, moral undertones at work in Jane's narrative in which she passively (hopelessly, one might say) desires a future that does not come to fruition and in her mourning of that lost possibility situates John as the agent of destruction and violation, furthering inscribing her own passivity.

However, again, it is important to note that, rather than being the result of disease (or a permanent pathological state of being), the forgetting that Jane has experienced is the result of the choice and desire to become intoxicated, which introduces the specter of morality. She has chosen to "become" temporarily or situationally pathological (intoxication to the point of incapacity). In shifting away from registering pathology as a state in which autonomy is not possible toward thinking about pathology within the frame of the situational, which is to say that it is regarded as a morally freighted choice, we see that situational incapacity is always followed by the shadow of autonomy.

Unlike Jane, John is positioned as criminal, not pathological. Here, we see that Jane's claiming of incapacitation frames John as the party who was not attentive to Jane and her inability to consent (despite his own intoxication). Occidental College's current policy on sexual assault is clear on this matter: "It shall not be a valid excuse to alleged lack of affirmative consent that the Respondent believed that the Complainant consented to the sexual activity under either of the following circumstances: (a) The Respondent's belief in affirmative consent arose from the intoxication or recklessness of the Respondent, or (b) the Respondent did not take reasonable steps, in the circumstances known to the Respondent at the time, to ascertain whether the Complainant affirmatively consented" (Occidental College 2016). John's intoxication retrospectively becomes scripted as dubious because it was a choice that was harmful to Jane in that he was not able to be attentive to her level of incapacity and misread her ability to consent. This shading of judgment has more to do with gender than memory in that it assumes that female chastity is sacrosanct and encodes a notion of women as more vulnerable and passive sexual agents into the law. Despite their presumptive equal degrees of incapacitation, John is still positioned as agential while Jane is assumed to have been passive.

In addition, John's ability to have intercourse was used as evidence against his claim of incapacity (Hess 2015). Through this framing of John as always agential and Jane as only occasionally so, we can see that discourses of incapacity meet consent in difficult ways when we think about temporality. The immediacy of John's erection was submitted as proof of not only his immediate desire, but of his ability to integrate that desire into a more coherent vision of selfhood. This frames him as a volitional actor and makes the situational legible as criminal rather than pathological. This is in sharp contrast to the distancing of Jane and Donna's carnal desires from any version of themselves, and it speaks to the gendered versions of incapacity at work. While Jane's in-the-moment actions conflict with assumptions about what women want, Donna's desires are dismissed because of her age and gender. Both Donna and Jane's incapacity falls under the rubric of pathology—Jane's situationally so and Donna's as a more permanent state.

These gendered framings of incapacity are what lead feminists such as Janet Halley to oppose affirmative consent. In outlining her concerns, Halley argues that affirmative consent laws "will enable people who enthusiastically participated in sex to deny it later and punish their partners. They will function as protective legislation that encourages weakness among those they protect. They will install traditional social norms of male responsibility and female helplessness" (2016, 259). What we see in Halley's protest against the moralism that underlies the enforcement of affirmative consent laws is that pathologization can be situational and temporary. We also see that while Fischel and O'Connell's model of sexual autonomy offers a way around pathology for those with disabilities, it does little to alter John and Jane's warring narratives.

In addition to showing the inadequacy of current models of consent to make sense of sexual interaction in which one or both parties is deemed incapacitated either temporarily or permanently, these cases illustrate the ways in which discourses of pathology can operate along gendered lines, fusing pathology with nonautonomy and furthering the assumption of female sexual passivity, which exists in contrast to (in these cases, criminalized) male sexual aggression. Thinking with these two models of the pathological (the situational and the state of being) means that consent is out of reach for people who have been deemed incapable by the courts—a group that includes people with severe intellectual and/or physical disabilities who cannot articulate *no* verbally or physically—and morally freighted for those who are temporarily incapacitated. Halley argues that one of the dangers of affirmative consent is that it enables people to retrospectively punish sexual partners, which we see happen in the case of Jane and John (2016). In grappling with situational pathology, discourses of consent also cannot make sense of multiple incapacitated parties—it requires that there be at least one volitional subject. Even in cases where pathology or incapacity is not at issue, consent often only underlines the inequality between both parties in the contract (Williams 1991).

How can we understand sexual desires and acts outside of this realm of nonconsent and pathology for those who have been deemed incapacitated? Here, I suggest that we dwell on the temporal difference between these frames of pathology. If situational pathology understands the aberrational moment as pathological because it produces the self as incoherent, how do we revalue the incoherence without framing it as pathological or criminal?

In the last section of this essay, I turn to an unlikely source, the journals of Gary Fisher, to think around consent and pathology to revalue aberrations of memory and the incoherence of selfhood. Fisher's journal entries provide an example of immediate desires, as recoded in his journal, as symptoms bound to his own past, the historical structures into which he was born, and his future desires.[3] In bringing together all of those temporalities, Fisher, a black gay male writer who died in 1993, shows the disconnect between selfhood and linearity and positions capacity in this space of temporal confusion. Consent drops out of the picture not only because Fisher is not writing from a place of forgetting either through intoxication or dementia, but because he understands himself not as an individual, but as part of a historical structure, and one cannot consent to be born into historical structures of oppression. This brings Fisher's narrative into conversations about incapacity, states of pathology, and consent most explicitly.

Fisher emphasizes the incoherence of his selfhood by dwelling on and eroticizing his position of structural inequality, thereby placing himself outside the bounds of the individual. In a journal entry, headed "November 2, 1985," he asks:

> Have I tried to oppress myself—as a black man and as a (passive) homosexual man—purely for the pleasure of it, or does that oppression go right to the point of my perceived weaknesses. It is societally impossible for me to make it the way I want to make it (financially, sexually, etc.)

> so I'm groping for an excuse, one that feels good and therefore must be good. Can I divorce sexuality from power in the real world or do I want to—here's one world *explaining* the other, and Christ, it's *so* hard to get answers. So I want to be a slave, a sex slave and a slave beneath another man's (a white man or a big man, preferably a big white man) power. Someone more aware of the game (and the reality of it) than myself. I want to relinquish responsibility and at the same time give up all power. I want to, in effect, give it to a system that wants to (has to) oppress me. (Fisher and Sedgwick 1996, 187)

Fisher's performance of and desire for black submission illustrates the complex relation between selfhood, agency, memory, and history. It also brings together the two different versions of pathology that I have been describing—Fisher is a pathological subject in that he is severed from discourses of autonomy due to his race (though arguably not incapacitated), and he enters into situational pathology by explicitly choosing to engage in oppressive situations. Fisher's insistence that his desire to submit both sexually and otherwise to a white man is inseparable from the history of black oppression in the United States stretches narrative memory beyond the parameters of the individual and toward the historical. Fisher's submission is a form of historical agency even as it reveals the limits of imagining individual subjectivity—his own narrative of self is inseparable from the temporality of generations and the historical time that cleaved blackness from humanity. Additionally, Fisher confesses that his actions run counter to "the real world's" expectations of the relationship between sexuality and history. Fisher should want to rebel, to be active, to want revenge; instead, he wants to exist in the temporality of the immediate—that of the sexual encounter—in order to dissolve these borders between the self and history. In relinquishing agency, autonomy, and selfhood, he does not give up eroticism or sex. Importantly, he enacts a temporal collapse in which capacity is not pure futurity, which we see in Jane's grief at the loss of her narrative of romance, but an amalgam with historicity. Fisher's capacity for transformation and movement is toward both a future and a past. Instead of the moralizing that happens when the future is too heavily weighted (John Doe) or the impasse that occurs when the past is privileged (Donna Rayhons), he illustrates the importance of thinking sex in the immediate in relation to both temporal orientations.

Further, in Fisher's insistence that sex register in the immediate (as well as the past and future), he points us toward an understanding of capacity vis-à-vis desire and constraint.

This is tricky terrain for theorizing consent and, indeed, consent's attachment to linear temporality, agency, and autonomy indicates that it may not be the appropriate rubric for assessing desire (even as it may allow for a discussion of violation). However, Fisher's journal entry exemplifies the multiplicity of memories that occur in the immediate, the difficulties of bounding off the self as individual, and the possibilities of capacity through attention to the immediacy of sex. Thinking with the immediate leads away from a narrative of cut-off possibilities and toward a theorization of the emergent and the possible as opposed to the idealized, wished-for. In the immediate, capacity becomes sensual. In the immediate we can register Donna's desire (and the desire of others who have been deemed pathological and thereby nonautonomous) on its own terms as something available and nonpernicious regardless of its narrative incoherence. In the immediate, we can also register the sensuality and desire at work behind moments when the self is incoherent (or the situationally pathological, as in the case of John and Jane Doe). Thinking with the immediate removes the framework of pathology and autonomy in order to think with the relational pleasures and possibilities of sensuality.

Notes

1 Henry Rayhons's case was widely discussed in the national media.

2 The *John Doe v. Occidental College* trial has also been widely chronicled.

3 At the time of his death, Fisher was a graduate student in English at University of California, Berkeley. When he died Fisher was unpublished, but through the efforts of his friends—most notably, the effort of Eve Kosofsky Sedgwick—his stories and his journal entries were published in a collected volume, *Gary in Your Pocket*. The back cover declares that the volume is "a uniquely intimate, unflinching testimony of the experience of a young African American gay man in the AIDS emergency."

Works Cited

Belluck, Pam. 2015. "Sex, Dementia and a Husband on Trial at Age 78." *New York Times*, April 13. Accessed June 29, 2016. www.nytimes.com/2015/04/14/health/sex-dementia-and-a-husband-henry-rayhons-on-trial-at-age-78.html?_r=0.

Brenoff, Ann. 2015. "Dementia And Sex: What Was Really on Trial with Henry Rayhons." *Huffington Post*, April 23. Updated June 23. Accessed June 29. www.huffingtonpost.com/ann-brenoff/dementia-and-sex-henry-rayhons_b_7122460.html.

Brudney, Daniel. 2009. "Beyond Autonomy and Best Interests." *Hastings Center Report* 39 (2): 31–37. doi:10.1353/hcr.0.0113.

California Senate. 2014. Senate Bill No. 967.

Dorment, Richard. 2015. "Occidental Justice: The Disastrous Fallout When Drunk Sex Meets Academic Bureaucracy." *Esquire*, March 25. www.esquire.com/news-politics/a33751/occidental-justice-case/.

Fischel, Joseph. 2013. "Against Nature, Against Consent: A Sexual Politics of Debility." *differences* 24 (1): 55–103. doi:10.1215/10407391-2140591.

Fischel, Joseph J., and Hilary R. O'Connell. 2016. "Disabling Consent, or Reconstructing Sexual Autonomy." *Columbia Journal of Gender and Law* 30 (2): 428–528.

Fisher, Gary, and Eve Kosofsky Sedgwick. 1996. *Gary in Your Pocket: Stories and Notebooks of Gary Fisher.* Durham: Duke University Press.

Halley, Janet. 2016. "Currents: Feminist Key Concepts and Controversies." *Signs: Journal Of Women in Culture and Society* 42 (1): 257–279.

Hasday, Jill Elaine. 2000. "Contest and Consent: A Legal History of Marital Rape." *California Law Review* 88 (5): 1373–1505. doi:10.2307/3481263.

Hess, Amanda. 2015. "How Drunk Is Too Drunk to Have Sex?" *Slate*, February 11. www.slate.com/articles/double_x/doublex/2015/02/drunk_sex_on_campus_universities_are_struggling_to_determine_when_intoxicated.html.

Kim, Scott Y.H. 2011. "The Ethics of Informed Consent in Alzheimer Disease Research." *Nature Reviews Neurology* 7: 410–414. doi:10.1038/nrneurol.2011.76.

Occidental College. 2016. "Sexual Respect & Title IX: Policies and Procedures." Occidental College. www.oxy.edu/sexual-respect-title-ix/policies-procedures.

Rubin, Gayle. 2002. "Thinking Sex: Notes for a Radical Theory of the Politics of Sexuality." In *Culture, Society and Sexuality: A Reader*, edited by Peter Aggleton and Richard Parker, 143–178. New York: Routledge.

Singai, Jesse. 2015. "Should Alzheimer's Patients Be Allowed to Have Sex?" *New York*, April 15, 2015.

Sulmasy, D.P., and L. Snyder. 2010. "Substituted Interests and Best Judgments: An Integrated Model of Surrogate Decision Making." *JAMA* 304 (17): 1946–1947. doi:10.1001/jama.2010.1595.

Williams, Patricia J. 1991. "On Being the Object of Property." In *The Alchemy of Race and Rights*, 216–238. Cambridge: Harvard University Press.

Yost, Rae. 2014. "Daughter Appointed Temporary Guardian of Rep. Rayhons' Wife." *Mason City Globe Gazette*, June 18. Accessed June 29, 2016. http://globegazette.com/news/local/daughter-appointed-temporary-guardian-of-rep-rayhons-wife/article_c7cc7709-cee5-5edf-913f-6110e73d2536.html.

Brains, Sex, and Queers 2090

An Ideal Experiment

ISABELLE DUSSAUGE

A Thought Experiment: Looking Forward to the Past

[. . .] In the 1985 film *Back to the Future* by Robert Zemeckis, the main character Marty McFly is accidentally sent to 1955 and spends the rest of the film going back and forth between present and past in order to secure that the future he wishes for 1955, that is, his parents starting a relationship, happens.[1] Borrowing the metaphor from Zemeckis' movie, I argue that much of my own work about the neurosciences, as other works in the same vein of criticism longing that things would be otherwise, functions as "going back" to undo what I wish had turned out differently so that something better could emerge instead. With such a gesture, what is to be undone is, for instance, dishonest and speculative neuroscience, sexist neuroscience, or neuroscience ignorant of feminist theory.

However, Zemeckis' movie points out that the gesture of going back to the future too often brings about major frustrations and dangers: First, the loop of going back and forth is difficult to stop. When attempting to undo undesirable pasts (or undesirable scientific accounts), I create new problems, which I attempt to undo again, etc.—hoping I do not meet myself on the way! Whereas it is not sure that these efforts succeed in

---

Isabelle Dussauge, "Brains, Sex, and Queers 2090: An Ideal Experiment," in Sigrid Schmitz and Grit Höppner (eds.), *Gendered Neurocultures: Feminist and Queer Perspectives on Current Brain Discourses*, pp. 67–88.

bringing forth a better present and future, it is certain that we critical researchers caught in this loop devote all our energy to that critical journey—which had better be worth the investment.

Second, going back to the future keeps bringing me back to the present again and leaves me with a sense of insufficiency. However necessary, unmaking problematic ex-futures does not substitute for fighting the struggles of the present and its own problematic futures. Neither does unmaking problematic science substitute for creating the alternative science I may wish for.

Third, in the many cases when going back to the future has not been sufficient to undo undesirable science, waking up in the morning to the same old problems for a day of undoing may feel a little like *Groundhog Day* (Murray, MacDowell, and Ramis 1993).

Consequently, the present chapter offers a temporary escape from this cyclic gesture of going back to the future, this scholarly life of undoings. Instead, I propose an attempt to *look forward to the past.*

Looking Forward to the Past

Looking forward to the past, that is, imagining the possible past as told from the perspective of a future moment, opens up for thinking historically about the present. When imagining the becoming past, one cannot miss the historical and sociological makings of knowledge, which too often seems to be the case in the critiques of the neurosciences which narrowly engage with neuroscientific knowledge on its own terms. Imagining the becoming past also helps avoiding the caveat, common to both neurosciences and the critical studies of the neurosciences, of rehearsing the pervasive modernist trope of the centrality of the experiment, which attributes to the scanner-experimental moment the power to produce the truth about the human and the world. Looking forward to the past is a thought experiment that demands of us critical researchers that we begin in the world outside the lab rather than in the scanner.

Now, let us fantasize about our future past for a moment.[2] In the following of this chapter, I invite you to imagine that we are gathered in September 2090 for a historical conference and that you are listening to a talk given at that conference. You may close your eyes and imagine that you know very little about what was going on in the 2000s and 2010s, in the world, in the neurosciences, and in the life that you usually consider as yours. Instead, the little you know about the early 2000s has been brought to you by history media (the everyday devices that have replaced history books).

You may now open your eyes and read the following as though you were listening, in 2090, to this conference about the past.

A Future History of Queer Feminist Neuroscience (2090)

Welcome to this historical lecture about the early days of queer feminist neuroscience in the wake of the 2000s. The following sequence is what is generally known about the beginnings of queer feminist neuroscience as we now know it.

> IN THE FIRST DECADES OF THE 2000s . . .
>
> Radical mobilizations happened in science.
>
> Women, queers, LGBT people grew tired of bad biological theories about their sexuality.
>
> If neuroscientists were to talk at all about sexuality, gender, or gender identity, they'd better get it right.
>
> Like their feminist predecessors, some of the queers believed that neuroscience would always be detrimental to them, and some of them believed that a new empirical agenda could be defined.
>
> (They acknowledged this conflict and all lived in peace.)
>
> Feminists in the late 20th century, and allied radicals, had proposed BOTH that neuroscience be criticized AND that feminist science be conducted.
>
> But no new feminist neuroscience had really emerged yet.
>
> The task was hard.
>
> But the wave of queer feminists was strongly determined to try again.

To continue this history, you understand that we cannot jump to parameters of neuroscientific experiments, as many historians and scientists too often do, unfortunately. We have to tell *why* these people went on, *who* did what, *how* it was possible and how it was organized—for *what purposes,* and doing *what.*

Why?

Twenty-first century neuroscience, just like twentieth-century science, reproduced heterosexist assumptions about the human, and had pervasive heterosexist effects in culture and society, especially in everyday sexual politics. This was well known by committed scientists and feminists after the work by second-wave feminists in the 1970s and onwards (Bleier 1984; Rose and Rose 1979), as well as renewed feminist scholarly work in the 2000s (Bluhm, Jacobson, and Maibom 2012; Fine 2012; Lancaster 2003). The critiques of the time often referred to this *de facto* alliance of neuroscience with sexism as "neurosexism," a term historians believe was coined by a young psychologist of the times, Cordelia Fine (2008, 2010).

Queers and feminists were frustrated because the neurosciences ignored their lives or defined them as anomaly from a norm.

Queers and feminists were also anxious: some of them because they believed neuroscience could yield the truth about themselves, others because they thought that neuroscience was widely popular and that people would generally believe what neuroscientists had to say about any topic, especially about gender and sexuality, which were salient political issues.

Moreover, in the increasingly technocratic, evidence-based European and American societies of the 2010s, minorities of power needed biological-scientific arguments to use in their social and political struggles for equity.

How?

To the best of our knowledge, the neuroscientific network *neuroQueer* was founded in Lausanne, Switzerland on May 10, 2014, during an event that took place during the conference "NeuroGenderings III." *NeuroQueer* gathered queer feminists: activists from different movements and scholars from different disciplines (neuroQueer 2014*).

They organized themselves along lines corresponding to what a sociologist of the time, Vololona Rabeharisoa had called a "participatory model" for social movements in medicine (2003). Rabeharisoa had been studying evidence based activism and she argued that a *participatory model* was characterized by an organization whose board worked to secure funding, retained control over social and scientific agendas, but were not conducing all the research themselves.

NeuroQueer's strategy was inspired by the French organization AFM *(Association Française contre les Myopathies),* which had been analyzed by Rabeharisoa but relied on the means of their own time.[3]

Where the AFM had been successfully raising funds year after year since the 1990s with a TV program called *Téléthon, neuroQueer* used crowdfunding from 2015 onwards. Crowdfunding—gathering funds on the internet directly from supporters and the general public—was a common strategy for activist and cultural projects in the 2010s. *NeuroQueer* organized cultural, political, and scientific events to give their agenda visibility and to generate funds. They also organized social activities: community building and community support; scientific activism and scientific empowerment. To define, build up, and run these social activities, they cooperated with LGBTQIA (lesbian/gay/bi/trans/queer/intersex/asexual) and feminist organizations, and with NGOs working on sexual rights and sexual health.

With the massive funding they gathered, they hired scientific groups through funded open calls for research. The board of *neuroQueer* organized the research and defined scientific priorities for the neuroscientific work of the organization. Those priorities were gender, sexuality, and power.

For What?

What did the *neuroQueer* network want? It is tempting—but perilous—to over-interpret the goals of *neuroQueer* in the light of their subsequent development and success. Complicating the matter further, most records from *neuroQueer's* archive were destroyed in the Great Radical-Conservative Attacks on NGOs and scientific journals in the 2040s. But if we go back to one of the few sources which remain from the network, minutes of a board meeting in February 2015, we find the following:

> neuroQueer KNOWLEDGE GOALS:
> - De-genitalize (and de-medicalize) sexuality! (following Tiefer 2002).
> - New cues about bodily sexual processes.

- How life & power leave marks on the brain & body (sexuality is a power order).
- Have fun with a scanner & see! (neuroQueer 2015*)

This document suggests that *neuroQueer* was inspired by the turn-of-the-century feminist sexology championed by Leonore Tiefer (2002) in the US, and by the feminist movements around the world, who, especially from the 1970s and onwards, had wanted to de-naturalize *and* de-medicalize sexuality. They also wanted to de-genitalize the physiology of sexuality.

For reasons which are not made explicit in this source, *neuroQueer* members wanted to find clues as to which trillions of human (individual and bodily) processes were going on in sexual situations. Moreover, and crucially, they wanted to study how lived life, sexual life, and power left possible marks on the body and brain. Finally, like many of the emergent neuroscientists of the time, they wanted to explore and test what could be known with functional neuroimaging methods.

The *neuroQueer* board also defined main principles for the directions of research, as the same document tells us:

> Guiding principles for problem selection:
>
> - Fair political economy of knowledge, benefiting (if at all) the communities upon which it is based.
> - End ongoing exploitation of LGBTI bodies, bodies of color and women's bodies for a better science for white cis-men.[4]
> - No neurocentrism: Why the brain? *What* in the brain? When do we need a scanner? (neuroQueer 2015*)

That document suggests that *neuroQueer* were aware that neurocentrism was part of the problem with mainstream neuroscientific knowledge, and with the neurocultures of the time: The late twentieth and early twenty-first centuries over emphasized and overvalued the brain in their understandings of the human. As cultural scientists such as Fernando Vidal and Francisco Ortega had documented, Western culture underwent a cerebralization in the late twentieth century, sustained by a somehow modernist belief that the brain contained the person and that the neurosciences could yield the ultimate truths about human behavior (Vidal 2009; Vidal and Ortega 2007, 2011). The rise of new neuro-disciplines such as neuropedagogy, neurolaw, and neurophilosphy was part of this cerebralization, and so were popular neurocultures such as "neurobics" (train-your-brain) movements and literature, and neuro-based Mars-and-Venus descriptions of gendered differences.

The minutes of the *neuroQueer* meeting in February 2015 (neuroQueer 2015*) also suggest that *neuroQueer* wanted to promote and produce a science which would, if at all, benefit their communities rather than the majority population. As several members of the network had documented—among others, gender and medical scientist Rebecca Jordan-Young (2010)—it had often been the case that trans, intersex, and LGB people had been used in neuroscientific studies and experiments to address problems in the science of gender more generally. To the best of our knowledge, the goal of that mainstream neuroscience of sex was never to make the lives of trans, intersex, LGB, or queer people better—the neuroscientists, most of them straight cis-persons, had usually no idea what these lives were or could be like. Instead, trans, intersex, and LGB people served as quasi-experimental instruments to distinguish gender from sexuality, or gender from gender identity, or gender related variables from one another in the neuroscience of sex (Jordan-Young 2010). In turn, the neurosciences of sex understood gender and sexuality as two binaries classifying the population as either "male" or "female," and either homo- or heterosexual, although these binaries did not and could not capture the life, body, or identity of any participant. (The thought that people naturally fell into these categories was a very popular belief in the twentieth and early twenty-first century!)

NeuroQueer's recognition of the unjust political economy of knowledge (Who benefits of the knowledge produced in whose bodies?) in the neurosciences served their emancipatory goal. Yet, the disposal of that unjust economy had a cost: the sacrifice of science as a universalist good. *NeuroQueer's* emancipatory goal demanded instead that knowledge should always be primarily aimed to benefit the minority group from which it was created, for instance women, queers, or trans people.

What would a neuroscience look like which embraced such aims, *neuroQueer* asked? Norm-critical queer feminist neuroscience of the time, such as in Anelis Kaiser's or Deboleena Roy's work (Dussauge and Kaiser 2012; Kaiser 2011, 2012; Kaiser et al. 2007; Kaiser et al. 2009; Roy 2007, 2012) may well have been laying the grounds for such an emancipatory goal, by taking the workings of power with the body and brain as its object of inquiry.

What?

The first publication with wide effects based on research funded by *neuroQueer* was the article "Power, Brains, and Sexuality," published in *Mature Neuroscience* (one of the most influential scientific journals of the time) in 2019 by researchers from gender studies, neurosciences, and sexology. (We have reasons to believe the article was published under pseudonyms—Elle G. Bitty, Queiras Volk, Q. Erfem Eneest, Kif H. Milk, and Harry Scott Snokstrom.)

"Power, Brains, and Sexuality" (PBS in the following) was a neuroscientific study of sexual acts. It was inspired by several of its contemporary or earlier directions of research: Firstly, by queer theoretical literature (1990s onwards), which in the early 2000s had shown a renewed interest in the political/emancipatory status of sexual desire and sexual practices—for instance, many scholars such as Judith Jack Halberstam (Halberstam 2008) and Fredrik Palm (Palm 2016)[5] had shown a critical interest in the earlier work of Leo Bersani (Bersani 1995, 1986) and Lee Edelman (Edelman 2004).

Secondly, PBS was clearly inspired by the work and proposals of several scholars in the network *NeuroGenderings*, many of whom also joined the *neuroQueer* network: Anelis Kaiser (Kaiser 2012; Dussauge and Kaiser 2012; Kaiser et al. 2009), Catherine Vidal (2012), Cordelia Fine (2010), Cynthia Kraus (2012a, 2012b), Deboleena Roy (2012), Emily Ngubia Kuria (2012), Hannah Fitsch (2012), Isabelle Dussauge (Dussauge and Kaiser 2013, 2012; Dussauge 2010), Katarina Hamberg (2010, 2000), Katrin Nikoleyczik (2012), Raffaella Rumiati (Jordan-Young and Rumiati 2012), Rebecca Jordan-Young (2010), Sigrid Schmitz (2012), and more.

Thirdly, PBS was building on the twentieth-century sexual sociology that was interested in sexual scripts (Kimmel 2007), however, in an updated theoretical "2.0" version which insisted on the performativity of socio-cultural scripts.

The *aims* of PBS were to "explore the productive relation of power attending to sexuality and brain" and to "highlight the temporal relations between lived sexuality (in power orderings of sexuality and gender) and the neural reflection of/participation in sexual desire and acts" (Bitty et al. 2019*, 962).

PBS therefore asked, "What sexual desires are possible, how do they unfold in the brain?" and inquired into the workings of power onto the neural involvements in sexual desire and sexual acts; and the neural involvements in desire, and their going-together in time with experiences of arousal.

PBS was based on a large-scale qualitative study. Participants were people of all genders with recurrent experience of occupying specific positions of minority of power (mostly: queers, women, LGBT, racified people; several of these positions). As controls, people were selected of all genders with no self-reported experience of minority of power (they were dominantly white, heterosexual cis-persons, a majority of them identifying as men).

Many innovations of PBS were *methodological*. Firstly, the participants were required to bring in objects (e.g., inanimate objects, outfits, bodily objects, music, erotica) or representations of situations (e.g., pictures/videos of actual objects, people, acts, music) towards which they often felt desire, or stories about those. They were also asked to bring related or similar kinds of objects, representations, or stories for which they usually felt no arousal—to be used as control situations. In the fMRI scanner, participants used the sexual situations they brought in, and the control situations.

Secondly, prior to the experiment, and between runs of the experiment, the subjects participated in semi-structured qualitative interviews about their sexual histories and current lives, their arousal/preferences in terms of object choice (other than gender, see Sedgwick 1990), their experiences of sexuality, and their experiences of troubled positions in and outside of sexuality (in line with Tiefer 2002).The semi-structured interviews conducted prior to the experiments were also used as a means to help participants to formulate possible objects and stories to bring in to the experiment. The interviews also served as a means for participants to make salient their experiences of positions of power, for instance, situations where they had experienced privilege or resistance.

During experiments, usually directly after a scanning session, participants were interviewed in order to formulate their first-person account of: experiences in the scanner; fantasies and meanings associated to arousal; relation between experience in the scanner, and outside-scanner sexual experiences. (For twenty-first-century reflections about first-person perspectives in the neurosciences, see Einstein 2012; Jack and Roepstorff 2003).

Thirdly, the temporal organization of the study reported in PBS was an innovation as compared to the short-sighted, for-profit studies of that time. The study repeated the neuroimaging experiments and qualitative interviews every three months for each participant in the group, for a period of two years. For a smaller random selection of twenty participants in both the subject group and control group, the imaging and interview sessions took place twice a month in the same period.

Fourthly, PBS' analytical work tested what they called "datacrowding" (yes, it was they who coined the term): The database of anonymized interviews, observations, and fMRI data were published electronically in open-access. PBS argued for its datacrowding model on scientific and ethical grounds, and it adopted an anti-dissemination stance, against the often corrupted bioethics of the time, which supported models of participation through better public scientific information. In the PBS study, data analysis was shared with queer feminist sexologists, psychologists, and activists, and other scientists and lay experts. The PBS team listened back to their interpretations of the data and confronted the results of divergent interpretations with one another, using scientific disagreements to clarify the lines of epistemological, social, and political conflicts (which in turn became part of the results as well). PBS thus aligned with the then nascent practice of dissensus work, championed by dissensus conferences, such as "NeuroGenderings III," which was held in Lausanne in 2014.

The open-access datacrowding was thus used to generate diverse results, interpretations, and debates, rather than closing them in advance. This was in line with how the PBS team envisaged scientific work: as a knowledge-generating organism which gathered experiences from people; made those meet scientific standards in order to deliver them to the scientific teams (this turned the participants into "experts on [their] experiences," Rabeharisoa 2003, 2133); invented new scientific categories when needed, for instance, for object choice and sexual preferences; and invented such categories in interaction with both participants and other relevant groups of experts.

In PBS, one level of the *analysis* was conducted with a renewed theoretical account of sexual scripts. The theory of "sexual scripts 2.0" (Bitty et al. 2019*, 963) acknowledged the sociological distinction between the societal and psychological levels of sexual scripts (Epstein 2007; Kimmel 2007) but insisted on a performative account of sexuality and gender (Butler 1990).

To the best of our knowledge, all experimental results, as well as the result part of the remaining publication, tragically disappeared together with *neuroQueer's* archive and *neuroQueer's* disseminated documents in the 2040s' Great Radical-Conservative Attacks on NGOs and scientific journals. Now—and this is a parenthesis—I can announce that my fellow historians of medicine may well have received funding to conduct a reconstruction of the later stages of the study, with the hope to understand better the context of the time, and to reconstruct hypothetical results of that seminal study. But this is still unofficial—and very exciting!

Now back to the late 2010s. PBS certainly became a historical landmark in the making of neuroscientific experiments, but first and foremost, it consecrated—and opened up—new ways to *organize* science and to conduct scientific activism.

"Power, Brains, and Sex" is one of the articles featured in the edited volume *Norm-Critical Queer Feminist Neuroscience* (neuroQueer 2020*), a programmatic book gathering seminal studies and their recursors conducted between 2015 and 2020.

Then . . . ?

However, some feminists, women, queers, and activists found it problematic to counteract neuroscience with more neuroscience. As the success of *neuroQueer's* activities became better known, they opposed *neuroQueer's* project either in its own right or for misdirecting political action. *NeuroQueer* acknowledged their resistance without trying to tame it. *NeuroQueer* seems to have held the view that a vibrant skepticism in their allied movements was required to keep skepticism strong in *neuroQueer* against neurocentrism or uncritical technological optimism.

This resistance from political allies to the *neuroQueer* project is illustrated by demonstrations (documented, for instance, in the daily paper *Feminist Global News)* featuring slogans such as "Stop medicalization: No science is innocent," "And Audre Lorde said: 'The master's tools . . .'," "Our bodies are for fun, not for science," "No scanner in my backroom," "Our bodies ourselves" (2022*).

Epilogue

NeuroQueer's *large-scale research program transformed the field of neuroscience. What is sure is that more and more people began to work with queer, feminist, norm-critical, non-neurocentric neuroscience. Concomitantly, more queers/women/LGBTs/people of color joined the field as first or second education, and some of them organized themselves as autonomous research groups and community science groups. As the bold research that was conducted in the* neuroQueer *network nudged mainstream neuroscience's understandings of gender, sexuality, and power, the new community science groups were able to get other public funds than* neuroQueer. *They also began to renew their scientific and political agenda in critical directions not foreseen by the original* neuroQueer *agenda. In 2035, after twenty years of existence, the* neuroQueer *research organization voted its own dissolution, on the grounds that community neuroscience had successfully emerged and taken over the vision of a radical neuroscience.*

Notes

1 *Editors' note:* See unabridged article for the sections titled "Epistemic Integrity" and "Renewing a Critical Gesture"

2 The remainder of this paper is an approximate retranscription *a posteriori* of the talk I gave at the conference. It has been a conscious choice to attempt to recreate the more informal tone of the speech and some of its visual materials.

3 Myalgia (English for *myopathie)* was a rare muscular disease without a cure, which is no longer the same since the therapeutic breakthroughs of the 2060s.

4 *Cis-person* was a term mainstreamed in gender-activist contexts in the two first decades of the 2000s and refers to a person in whom biological gender and gender identity are aligned. Roughly speaking, a cis-person denoted the position of not being a trans person. Activists used the term cis-person in order to make visible the privileges which were associated with the gender norm, and which trans people were denied. The origin of the term is unclear.

5 *Editors' note:* Palm's article was not yet published when this piece was first printed; the citation has been updated.

Archives

nQA *neuroQueer* Archive, Lausanne, Switzerland, 2012–2035.
DPA Isabelle Dussauge's personal archive, Stockholm, Sweden, 1999–2064.

Works Cited

Fake references are indicated with an asterisk *.

2022*. "Will Feminists Reject Neuroscience?" *Feminist Global News*, 8 March.

Bersani, Leo. 1986. *The Freudian Body: Psychoanalysis and Art*. New York: Columbia University Press.

———. 1995. *Homos*. Cambridge, MA: Harvard University Press.

Bitty, Elle G., Queiras Volk, Q. Erfem Eneest, Kif H. Milk, and Harry Scott Snokstrom. 2019*. "Power, Brains, and Sexuality." *Mature Neuroscience* 20 (7): 962–70.

Bleier, Ruth. 1984. *Science and Gender: A Critique of Biology and Its Theories on Women, Athene Series*. New York: Pergamon Press.

Bluhm, Robyn, Anne Jaap Jacobson, and Heidi Lene Maibom, eds. 2012. *Neurofeminism: Issues at the Intersection of Feminist Theory and Cognitive Science*. New York: Palgrave Macmillan.

Butler, Judith. 1990. *Gender Trouble: Feminism and the Subversion of Identity*: Psychology Press.

Dussauge, Isabelle. 2010. "Sex, Cash, and Neuromodels of Desire." Neurosociety: What Is It With the Brain These Days?, Oxford, 7–8 December 2010.

Dussauge, Isabelle, and Anelis Kaiser. 2012. "Re-queering the Brain." *Neurofeminism: Issues at the Intersection of Feminist Theory and Cognitive Science*: 121–144.

———. 2013. "Repolitisations féministes et queer du cerveau." *Revue d'anthropologie des connaissances* 7 (3): 667–692.

Edelman, Lee. 2004. *No Future: Queer theory and the Death Drive*. Durham: Duke University Press.

Einstein, Gillian. 2012. "Situated Neuroscience: Exploring Biologies of Diversity." In *Neurofeminism: Issues at the Intersection of Feminist Theory and Cognitive Science*, edited by Robyn Bluhm, Anne Jaap Jacobson and Heidi Lene Maibom, 145–74. New York: Palgrave Macmillan.

Epstein, Steven. 2007. "'The Badlands of Desire': Sex Research, Cultural Scenarios, and the Politics of Knowledge Production." In *The Sexual Self: The Construction of Sexual Scripts*, edited by Michael S. Kimmel, 249–63. Nashville: Vanderbilt University Press.

Fine, Cordelia. 2008. "Will Working Mothers' Brains Explode? The Popular New Genre of Neurosexism." *Neuroethics* 1 (1): 69–72.

———. 2010. *Delusions of Gender: How Our Minds, Society, and Neurosexism Create Difference*: WW Norton & Company.

———. 2012. "Explaining, or Sustaining, the Status Quo? The Potentially Self-Fulfilling Effects of 'Hardwired' Accounts of Sex Differences." *Neuroethics* 5 (3): 285–94.

Fitsch, Hannah. 2012. "(A)e(s)th(et)ics of Brain Imaging. Visibilities and Sayabilities in Functional Magnetic Resonance Imaging." *Neuroethics* 5: 275–83. doi: 10.1007/s12152-011-9139-z.

Halberstam, Judith. 2008. "The Anti-social Turn in Queer Studies." *Graduate Journal of Social Science* 5 (2): 140–156.

Hamberg, Katarina. 2000. "Könet i hjärnan." *Läkartidningen* 45: 5130–36.

———. 2010. "Understanding Sex or Gender Differences in Neuroscience—Stressing the Context and Avoiding Essentialism." Neurogenderings: Critical Studies of the Sexed Brain, Uppsala, Sweden, 25–27 March.

Jack, Anthony I, and Andreas Roepstorff. 2003. *Trusting the Subject? Volume 1 the Use of Introspective Evidence in Cognitive Science*. Imprint Academic.

Jordan-Young, Rebecca. 2010. *Brain Storm: The Flaws in the Science of Sex Differences*. Cambridge, MA: Harvard University Press.

Jordan-Young, Rebecca, and Raffaella I. Rumiati. 2012. "Hardwired for Sexism? Approaches to Sex/Gender in Neuroscience." *Neuroethics* 5: 305–315. doi: 10.1007/s12152-011-9134-4.

Kaiser, Anelis. 2011. "Neurofeminism and the Cortical Power of Gender Differences." Breaking the Norms, Frankfurt am Main, 6–7 October.

———. 2012. "Re-conceptualizing 'Sex' and 'Gender' in the Human Brain." *Journal of Psychology* 220 (2): 130–6.

Kaiser, Anelis, Sven Haller, Sigrid Schmitz, and Cordula Nitsch. 2009. "On Sex/Gender Related Similarities and Differences in fMRI Language Research." *Brain Research Reviews* 61 (2): 49–59.

Kaiser, Anelis, Esther Kuenzli, Daniela Zappatore, and Cordula Nitsch. 2007. "On Females' Lateral and Males' Bilateral Activation During Language Production: A fMRI Study." *International Journal of Psychophysiology* 63 (2): 192–198.

Kimmel, Michael S. 2007. *The Sexual Self: The Construction of Sexual Scripts*. 1st ed. Nashville: Vanderbilt University Press.

Kraus, Cynthia. 2012a. "Critical Studies of the Sexed Brain: A Critique of What and for Whom?" *Neuroethics* 5: 247–259. doi: 10.1007/s12152-011-9107-7.

———. 2012b. "Linking Neuroscience, Medicine, Gender and Society through Controversy and Conflict Analysis: A 'Dissensus Framework' for Feminist/Queer Brain Science Studies." In *Neurofeminism: Issues at the Intersection of Feminist Theory and Cognitive Neurosoence*, edited by Robyn Bluhm, Anne Jaap Jacobson and Heidi Maibom, 193–215. Basingstoke: Palgrave Macmillan.

Kuria, Emily Ngubia 2012. "The Challenge of Gender Research in Neuroscience." In *Essays on Neuroscience and Political Theory: Thinking the Body Politic*, edited by Frank Vander Valk, 268–87. New York: Routledge.

Lancaster, Roger N. 2003. *The Trouble with Nature: Sex in Science and Popular Culture*. 1 ed: University of California Press.

Murray, Bill, Andie MacDowell, and Harold Ramis. 1993. Groundhog Day. Burbank, CA: Columbia Pictures.
neuroQueer. 2014*. "Poster conference " neuroQueer, Lausanne, Switzerland, nQA.
———. 2015*. neuroQueer minutes. Lausanne, Switzerland, nQA.
———. 2020*. *Norm-Critical Queer Feminist Neuroscience*. London: Rootledge.
Nikoleyczik, Katrin. 2012. "Towards Diffractive Transdisciplinarity: Integrating Gender Knowledge into the Practice of Neuroscientific Research." *Neuroethics* 5: 231–245. doi: 10.1007/s12152-011-9135-3.
Palm, Fredrik. 2016. "Sexual Arousal, Danger, and Vulnerability." In *Bodies, Boundaries and Vulnerabilities*, 119–140. Springer.
Rabeharisoa, Vololona. 2003. "The Struggle Against Neuromuscular Diseases in France and the Emergence of the 'Partnership Model' of Patient Organisation." *Social Science & Medicine* 57 (11): 2127–2136.
Rose, Hilary, and Steven Rose. 1979. "Radical Science and Its Enemies." *Socialist Register,* 16: 317–35.
Roy, Deboleena. 2007. "Somatic Matters: Becoming Molecular in Nolecular Biology." *Special Issue, Rhizomes: Cultural Studies in Emerging Knowledge* 14.
———. 2012. "Neuroethics, Gender and the Response to Difference." *Neuroethics* 5: 217–230. doi: 10.1007/s12152-011-9130-8.
Schmitz, Sigrid. 2012. "The Neurotechnological Cerebral Subject: Persistence of Implicit and Explicit Gender Norms in a Network of Change." *Neuroethics* 5:261–274. doi: 10.1007/s12152-011-9129-1.
Sedgwick, Eve Kosofsky. 1990. *Epistemology of the Closet*: University of California Press.
Tiefer, Leonore. 2002. "Arriving at a 'New View' of Women's Sexual Problems: Background, Theory, and Activism." *Women & Therapy* 24 (1–2): 63–98.
Vidal, Catherine. 2012. "The Sexed Brain: Between Science and Ideology." *Neuroethics* 5: 295–303. doi: 10.1007/s12152-011-9121-9.
Vidal, Fernando. 2009. "Brainhood, Anthropological Figure of Modernity." *History of the Human Sciences* 22 (1): 5–36.
Vidal, Fernando, and Francisco Ortega. 2007. "Mapping the Cerebral Subject in Contemporary Culture." *Elect. J. Commun. Inf. Innov. Health* 1: 255–259.
———. 2011. "Approaching the Neurocultural Spectrum: An Introduction." *Vidal F, Ortega F. Neurocultures. Glimpses into an Expanding Universe, Frankfurt am Main: Peter Lang.*
Zemeckis, Robert. 1985. *Back to the Future*. Universal City, CA: Universal.

PART FOUR

Beyond the Human

IMAGINE, FOR A MOMENT, THAT WE, WHOMEVER THAT IS IMAGINED to be, have managed to "stop telling ourselves the same old anthropocentric bedtime stories" (Shaviro 1997 via Barad 2003). Perhaps a place where, by necessity (first contact) or attrition (every other path didn't work), the primacy of one (white) (male) (cis) (colonizer) (straight) (abled) (human) (animate) experience has given way to a holistic view of the universe, where the boundaries of sex/gender and sexuality are widely accepted to be permeable—even, in some contexts, to have outlived usefulness. Perhaps you imagine a time where our current prejudices fall out only to be replaced with new boundaries: The rules of who plays may change, but the game of exclusion remains. Your experience of feminist queer futurity may vary (Muñoz 2009; Edelman 2004; Seymour 2013), but whatever it may be, we invite you, in the spirit of Dussauge's thinking forward to the past ("Brains, Sex, and Queers 2090: An Ideal Experiment"), to sit for a while in this (queer) time and place (Halberstam 2005), and think on both its potential and historical roots. What ontological and epistemological possibilities exist once we move beyond the twin presumptions that all subjects are human and that the measure of nature is objective? Whose work is foundational? What ideas do its scholars see as groundbreaking? Whose science persists, whose is abandoned, and whose is born? These are the questions central to the fourth and final section of our volume, where we explore queer feminist science studies at its edges, engage the speculative, and construct our own u-/dystopian genealogies.

As a whole, queer feminist work beyond the human is varied, vast, wide-ranging, and delightfully esoteric. It emerges in the examination of relationships between human and nonhuman animals, pointing out the fundamental disconnect of seeking confirmation of ourselves in the behavior of animals whilst also asserting our own primacy over nature or examining speciality, companionship, coevolution, and the varieties (and precarities) of biodiversity (Haraway 2003, 2008; Roughgarden 2004). Queer and feminist theory also claims spaces in the nonhuman; in examinations of monsters, perverts, and other (in)humanities (MacCormack 2009), engagements with nomadic embodiment, cyberculture, and the posthuman (Braidotti 2006, 1994; Hayles 1999; Halberstam and Livingston 1995), and the emergence of relations between the inanimate and the animate (Terry 2010). And finally, queer feminist science studies beyond the human directly engages the realms of speculative fiction, drawing on a rich history of cross-pollination between speculative fiction and social justice (Imarisha and brown 2015). These immersive explorations of reproductive dystopias (Atwood 1986), xenobiological relationships to sex and gender (Le Guin 1969; Butler [1987] 1997), and alternate gender realities (Brantenberg 1985) serve not only as political commentary but also as articulations of queer feminist possibility. The five pieces included in this section represent these forms of engagement—alternately speculative and relative, natural and unnatural—as we engage the questions of beyond.

Part Four begins with Angela Willey and Sara Giordano's essay "'Why Do Voles Fall in Love?': Sexual Dimorphism and Monogamy Gene Research," where they interrogate how researchers transpose cultural assumptions about gender onto animals (in this case, voles) in an ultimate search for "what makes the human human." Willey and Giordano question the presumption of sexual dimorphism in monogamy gene research in order to rethink how biologizing discourses naturalize anthropocentric, heteronormative narratives of coupling and belonging. In so doing, they call us to consider the questions we might ask if sex, gender, and sexuality were not pretheoretical assumptions upon which our conceptions of (human and nonhuman) nature were founded.

In her essay "Natural Convers(at)ions: Or, What if Culture Was Really Nature All Along?" Vicki Kirby throws a queer, postmodern, and material wrench into the nature/culture debates by introducing a radical new framework for the significance of nature. In this essay, first published in the collection *Material Feminisms* (Alaimo and Hekman 2008), Kirby brings older so-called "essentialist" feminism, one in which women were allied with nature (Ortner 1972), into conversation with a tension she sees in current feminist and science studies articulations of nature/culture: namely, that these interrogations of the nature/culture divide purport to undermine the entire apparatus of critique built on this distinction, and yet remain skeptical, and even distrustful, of alliance with nature. What would happen, she concludes, if feminism, instead, embraced the category of the natural?

In the selections from her 2004 book *Abstract Sex*, Luciana Parisi explores what it means to have sex without bodies. In "The Bio-technological Impact," Parisi points out that Haraway's dream of the postgender cyborg (1985) has still failed to materialize (the editors apologize for this pun), and attributes this to the ultimate failure of the displacement of the signifier of sex/body. Postgender feminism, Parisi argues, is trapped by the choice between embodiment ("biological essentialism") and disembodiment ("discursive constructivism"). She proposes abstract sex as a third alternative—one that relies on a radical, ground-up redefinition of bodies and matters. In the next selection, "Abstract Sex," Parisi draws on the work of Deleuze and Guattari to define abstract matter as machinic and mutating (Deleuze 1972, 1988a, 1988b, 1990, 1994; Guattari 1984; Deleuze and Guattari 1988). Abstract sex, as the abstract matter machine produced by joining previously separated abstract matter machines, becomes a capacious category.

In "Toxic Animacies, Inanimate Affections," Mel Chen takes a recent flurry of panic over lead-based paint in children's toy trains as a entry point into a discussion of the entanglement of race, gender, and sexuality in phobias about invasion and toxicity and, ultimately, as a gate toward opening up a space for what Chen calls "the queer productivity of toxins." Chen shows how the fear about lead-based paint on toys is a fear of invasion of the other into the home (the unsafe Chinese-made toy into the safe suburban American home), and of the other into the body (the young boy perversely licking the train), fears deeply entwined with questions of race, class, labor, status, and nationalism. To think with toxicity, Chen says, necessitates thinking with animacy, as it imbues the toxic object with affective strength. Underlying our fear of toxins is knowledge that the boundary between the inanimate and the animate, between the toxic and the safe, between the self and the other, is a permeable one.

In the final original contribution, "Plasmodial Improprieties: Octavia E. Butler, Slime Molds, and Imagining a Femi-Queer Commons," Aimee Bahng explores the possibilities of speculative fabulation, positing science fiction author Octavia Butler as a black feminist philosopher of science whose work examines the potentials of nonhuman engagement while also considering the "histories of empire and slavery as phenomena at the planetary scale." First, Bahng reads Butler's engagement with slime molds alongside those of feminist biologist Evelyn Fox Keller, noting both as moments where the slime mold disrupts current understanding of hierarchical developmental systems, whether they be within civilizations or single organisms. But, Bahng reminds us, through readings of Butler's Xenogenesis and Earthseed trilogies, this is "no utopia." "There is reason to pause the celebration of the liberatory potential of the social amoebae," she says. "Innovators and entrepreneurs have folded slime molds into the workforce as experimental bodies, picked up for their efficiency and utility, but not for their queerness." Bahng's work brings our reader to a close with grounded imagination: She invites us to consider speculative fabulation as a queer feminist science studies methodology where the queerly troubled future is guided by a critical eye.

Discussion Questions

1. Each of these essays figures the human and/or nonhuman a bit differently. What commonalities or continuities and divergences do you see in their approaches?
2. What challenges do these essays make to foundational definitions? What are these challenges meant to do?

Works Cited

Alaimo, Stacy, and Susan J. Hekman. 2008. *Material Feminisms*. Bloomington: Indiana University Press.

Atwood, Margaret. 1986. *The Handmaid's Tale*. Boston: Houghton Mifflin.

Barad, Karen. 2003. "Posthumanist Performativity: Toward an Understanding of How Matter Comes to Matter." *Signs* 28 (3): 801–831.

Braidotti, Rosi. 1994. *Nomadic Subjects: Embodiment and Sexual Difference in Contemporary Feminist Theory, Gender and Culture*. New York: Columbia University Press.

———. 2006. *Transpositions: On Nomadic Ethics*. Cambridge, UK: Polity.

Brantenberg, Gerd. 1985. *Egalia's Daughters: A Satire of the Sexes*. US ed. Seattle: Seal Press.

Butler, Octavia E. (1987) 1997. *Dawn*. New York: Warner Books.

Deleuze, Gilles. 1972. *Proust and Signs*. New York: G. Braziller.

———. 1988a. *Bergsonism*. New York: Zone Books.

———. 1988b. *Spinoza: Practical Philosophy*. San Francisco: City Lights Books.
———. 1990. *The Logic of Sense*. New York: Columbia University Press.
———. 1994. *Difference and Repetition*. London: Athlone Press.
Deleuze, Gilles, and Félix Guattari. 1988. *A Thousand Plateaus: Capitalism and Schizophrenia*. Translated by Brian Massumi. London: Athlone Press.
Edelman, Lee. 2004. *No Future: Queer Theory and the Death Drive*. Durham: Duke University Press.
Guattari, Félix. 1984. *Molecular Revolution: Psychiatry and Politics*. New York: Penguin.
Halberstam, Judith. 2005. *In a Queer Time and Place: Transgender Bodies, Subcultural Lives*. New York: NYU Press.
Halberstam, Judith, and Ira Livingston. 1995. *Posthuman Bodies*. Bloomington: Indiana University Press.
Haraway, Donna, 1985. "A Manifesto for Cyborgs: Science, Technology, and Socialist Feminism in the 1980s." *Socialist Review* 80 (March–April 1985): 65–107.
———. 2003. *The Companion Species Manifesto: Dogs, People, and Significant Otherness*. Chicago: Prickly Paradigm Press.
———. 2008. *When Species Meet*. Minneapolis: University of Minnesota Press.
Hayles, N. Katherine. 1999. *How We Became Posthuman: Virtual Bodies in Cybernetics, Literature, and Informatics*. Chicago: University of Chicago Press.
Imarisha, Walidah, and adrienne maree brown, eds. 2015. *Octavia's Brood: Science Fiction Stories from Social Justice Movements*. Chico: AK Press.
Le Guin, Ursula K. 1969. *The Left Hand of Darkness*. New York: Walker.
MacCormack, Patricia. 2009. "Queer Posthumanism: Cyborgs, Animals, Monsters, Perverts." In *The Ashgate Research Companion to Queer Theory*, edited by Noreen Giffney and Michael O'Rourke, 111–26. Farnham, UK: Ashgate.
Muñoz, José Esteban. 2009. *Cruising Utopia: The Then and There of Queer Futurity*. New York: NYU Press.
Ortner, Sherry B. 1972. "Is Female to Male as Nature Is to Culture?" *Feminist Studies* 1 (2): 5–31.
Parisi, Luciana. 2004. *Abstract Sex: Philosophy, Bio-technology and the Mutations of Desire*. New York: Continuum.
Roughgarden, Joan. 2004. *Evolution's Rainbow: Diversity, Gender, and Sexuality in Nature and People*. Berkeley: University of California Press.
Seymour, Nicole. 2013. *Strange Natures: Futurity, Empathy, and the Queer Ecological Imagination*. Urbana: University of Illinois Press.
Shaviro, Steven. 1997. *Doom Patrols: A Theoretical Fiction about Postmodernism*. London: Serpent's Tail.
Terry, Jennifer. 2010. "Loving Objects." *Trans-Humanities* 2 (1): 33–75.

"Why Do Voles Fall in Love?"

Sexual Dimorphism in Monogamy Gene Research

ANGELA WILLEY AND SARA GIORDANO

> Once upon a time, there was a meadow vole who was quite promiscuous in his behavior. He would mate with several voles and practically ignore his children. His cousin, the prairie vole, on the other hand, remained faithful to one female vole. So, scientists decided to give extra vasopressin (a hormone found in the prairie vole) receptors to the meadow voles, which have fewer vasopressin receptors. "The results were remarkable. After the V1a receptor gene was introduced, the former playboys reformed their ways. Suddenly, they fixated on one female, choosing to mate with only her—even when other females tried to tempt them," reported the BBC News.
>
> ANDREA AND ALICIA, *THE NEW VIEW ON SEX*

IN RECENT YEARS THERE HAS BEEN MUCH TALK IN THE POPULAR press and in popular culture about monogamy and animals (e.g., Jacquet 2005). Voles have become popular in recent monogamy research because they are reported to form monogamous relationships or to be promiscuous, depending on their species. As the story above—told by chastity educators with Pregnancy Center East in Cincinnati, Ohio, on their blog, *The New View on Sex* (Alicia 2008)—explains, prairie voles are considered a monogamous species, while the meadow vole is said not to pair-bond. Vole research on "the monogamy gene" since 2004 has received mention in the *New York Times*, *The Nation*, *Al Jazeera*, and various local papers as well as on *Late Night with David Letterman*, *The Daily Show*, *Dateline on NBC*, and National Public Radio. Some headlines focusing on genetic links to monogamy in voles read: "To Have and to Vole" (Ballon 2005), "How Geneticists Put the Romance Back into Mating" (Johnston 2005) and "Love Is a Drug for Prairie Voles to Score" (Sample 2005). A popular education module put out on YouTube by the laboratory on whose work we focus boasts a similarly provocative title: "Why Do Voles Fall in Love?"[1]

This is a play, of course, on the title of the Frankie Lymon and the Teenagers 1956, oft-covered hit song, "Why Do Fools Fall in Love?" The song made *Rolling Stone*'s top five hundred songs of all time list (November 2004); it is a deeply culturally entrenched question. The insights into the mystery that vole research on monogamy purports to offer, we argue, are similarly overdetermined. The story of "love"—in science and the larger culture of which it is a part—is clearly a gendered one. Discursively speaking, male monogamy and female monogamy both are and are not the same phenomenon to be explained. On the one hand, "monogamy" is gender neutral—it is about pair-bond formation, attaching to a mate. Both presumably heterosexual partners would need to exhibit a particular set of behaviors for a bond to be said to exist. These criteria include some form of cohabitation, mutual grooming, and co-parenting. As our introductory quote makes plain, female monogamy is more or less self-evident. It is taken for granted that a female who has mated and borne offspring is attached to her offspring's genetic other-half. We are given a fairly explicit genetic account of male mating behaviors, however, one that offers up two readings that have both been exploited in press coverage of this research and captured concisely in a cartoon from the online magazine *Science and Spirit* (Snider 2002) (see following page). First, there is a sense of the malleability of the genetic self. Second, and implicit in the first, is the naturalization of male infidelity. This is coupled with an implicit damnation of female promiscuity, for which no genetic explanation is on offer.

Precisely because of its socially monogamous nature, "the prairie vole (*Microtus ochrogaster*) has emerged as one of the preeminent animal models for elucidating the genetic and neurobiological mechanisms governing complex social behavior" (Young Lab). Because the prairie vole, unlike rats and mice, is said to be monogamous, it is fast becoming "one of the most powerful animal models for basic and translational research with direct implications for human mental health" (Young Lab). In her study of the standardization of animals for experimentation in the first half of the twentieth century, Karen Rader emphasizes how cultural assumptions about animals inform their use as models in a given sociohistorical context (1995, 252). The move to using prairie voles for the modeling of human

"Enhanced Vasopressin Receptor" cartoon. © 2004 Clifford Jay Snider. All rights reserved.

behaviors marks the consolidation of an implicit cultural consensus about monogamy as somehow fundamental to what makes the human, human. We are concerned about the implications of this formulation, especially for women for whom monogamy historically has warranted some explanation. We are thinking here of images of black women as sexually voracious (Hill Collins 2004), Latinas as hypersexual (Arrizón 2008), masculine lesbians as sexually predatory (Hantzis and Lehr 1994), and poor women, often racialized, as promiscuous mothers of too many children (Hill Collins 1990).

As a feminist neuroscientist interested in ethics broadly construed (Sara Giordano) and a feminist theorist and historian of sexuality in science interested in monogamy's role in the production of normal and abnormal bodies (Angela Willey), we are interested in different aspects of this research and its representation in the press. Given the privileged epistemic status of science in our culture, we are both invested in understanding how assumptions about gender and difference more broadly inform its truth claims. We share a strong interest in the role that assumptions about sexual dimorphism play in the production of "a monogamy gene." Dividing populations of living things into binary sex categories for the purposes of elucidating the workings of the natural world has come to seem so logical, so obvious. Even as the efficacy of those categories is challenged within and outside of the laboratory, neuroscientific research, particularly as it relates to "love," persists in pursuing research agendas premised on the idea that physiological processes are consistent enough to warrant generalization within a sex category while being wholly different across sexes. [. . .]

Destabilizing Sex

[. . .] The study of sexual strategies in general, not just monogamy, has played a role in the naturalization of male rape, female caregiving, and heterosexuality itself (Hubbard 1990). The importance of these sexual strategies hinges on the evolutionary assumption that each individual organism (human or otherwise) has as its primary "goal" the perpetuation of its own genetic material through reproduction. The differential strategies that scientists look for in women and men are linked to scientific representation of gametes: sperm are plentiful and mobile, hence males/men optimize their chances of reproduction by fucking everything. Eggs are seen as both stationary and finite in number, so females/women maximize their genetic survival by selectively choosing how to make the most of their seed. In fact, as dissenting scientific voices since at least 1948 have pointed out, this is but one way of interpreting the "lives" of the egg and the sperm. Contrary to the active/passive tale of fertilization, two cells fuse (active/active) and sperm are not as strong, goal-oriented, or potent as some scientists have imagined (Martin 1991).[2]

Assumptions about the gendered structure of pair bonding are rarely challenged as it passes unexamined as "natural." "Human nature" is "a normative concept that incarnates (in the literal sense of enveloping in flesh) historically based beliefs about how people should behave"

(Hubbard 1990, 107). These beliefs are strongly steeped in ideas about gender difference and the gender division of labor. Ruth Hubbard argues that women's biology, the sexed body as it were, is political, in part because the social and biological cannot be separated. [. . .] Group differences cannot be generalized and attributed to biology in a society where different groups are treated differently and have differential access to resources. In terms of monogamy, there may well be evidence—perhaps even physiological evidence—that women desire monogamous relationships. This cannot readily be disentangled from the many compelling explanations feminists have offered for the existence of compulsory heterosexuality (Rich 1980) and compulsory monogamy (Emens 2004; Murray 1995; Rosa 1994). [. . .]

At different moments (and certainly in different places), certain formations and combinations of aspects of "biological sex" are seen as essential or incidental or somewhere in between. Still, scientific research proceeds from subject recruitment and selection to the reporting of findings as if sex were binary, timeless, and self-evident. In terms of monogamy gene research, the sexing of voles is nowhere explained and the dimorphic and gendered structure of pairbond formation is largely taken for granted. From Willey's interviews we discovered the pragmatics of vole sex assignment. The sole criterion is the distance between the anus and the dot that is their genitals. This measurement is approximate and imprecise. The only check on accuracy is that the animals are sex segregated in their cages and, if they are kept alive long enough, will sometimes reproduce, so the animals in that cage will have to be re-sexed.

With an understanding of the notion of sex as we know it as a historical process, we go on to explore how these stories about sex and sex difference show up in primary science literature. We begin with a discussion of the V1aR gene, "the monogamy gene."

A Gene for Monogamy?

In our introduction we highlighted news reports on the discovery of the monogamy gene. Although the idea of a "gene for" a particular condition or attribute is widely understood as misleading, the concept prevails in popular and scientific cultures (Hubbard and Wald 1999; Lewontin 1991). The idea of the gene preceded its materiality, that is, the discovery of DNA.

The sciences and politics of eugenics were based on the concept of heritability that was fixed in the body, although a physical location of the gene did not yet exist (Hubbard and Wald 1999; Ordover 2003). The most common meaning of "gene" remains "something that is heritable." The more technical and more recent scientific definition of a gene is a region of DNA that "codes" for a specific protein. In order for a protein to be synthesized, many other proteins must be involved. Whether, how, when, and where a protein is produced, as well as how much is produced, is highly dependent on numerous factors outside the gene itself (Hubbard and Wald 1999). Although we tend to think of genes as deterministic of behavior, there are many steps to consider between a region of DNA that has something to do with the production of a protein to how that protein interacts with other proteins and how those interactions affect a specific behavior. [. . .]

Many assumptions and simplifications must be made to make the association between behaviors understood as monogamous and the basic genetic research that is the impetus for such news reports. Many of the reports of a gene for monogamy are based on Elizabeth Hammock and Larry Young's (2005) research, in which voles were bred to have varying lengths of a specific region of DNA. The region of interest is not part of what would technically be considered a gene—it is not in the "coding" region but rather in an area surrounding the gene for a vasopressin hormone receptor, V1aR, in voles. Researchers believe that this region affects the production of these vasopressin receptors in certain brain areas. Regions of DNA outside of coding regions were until recently commonly regarded as useless stretches and often referred to as "junk DNA." Now these regions are increasingly being examined as potentially important in the regulation of protein synthesis. The voles were grouped as having either shorter or longer than average lengths of these stretches of DNA. Then these groups were compared through a series of quantifiable behavioral tests that are used as proxies for more complex behaviors such as parenting and partnering. So, for example, one result that the researchers reported was that the group with longer lengths licked their pups more often than the group with shorter lengths. These were group differences, so once again it is important to remember that if you compared a vole with a short length to one with a long length, there may be no difference in the number of times they licked their pups or the result might be the opposite of the group result. [. . .]

Studies of vasopressin and the V1a receptor have been conducted almost exclusively in male animals. Most of the research from Young's lab on pair bonding has been on vasopressin receptors in the brains of voles. Vasopressin is widely known for its regulation of kidney function and is also used in regulation of blood pressure in humans. These peripheral effects have not been described as sex-specific. However, for its role in pair bonding or monogamy, it has been specifically investigated and linked to male behaviors. Although reports had failed to find sex-specific distribution of vasopressin receptors in both wild and lab-bred populations (Phelps and Young 2003), the search for sexual dimorphism in this system has continued. The vasopressin systems have recently been described as three separate systems where in two of the systems there is no sexual dimorphism but in the third there is some evidence of receptor distribution differences (Lim, Hammock, and Young 2004). Tracing research on the V1a receptor tells us that the pursuit of sexually dimorphic processes has shaped research on what has become "the monogamy gene" at every stage. In the next section we will examine the place of female sexuality in this research.

Sexed Research Agendas

Having problematized the idea of a "gene for" and the binary nature of "sex," we move from the theory to the laboratory and from the general to the specific. In Willey's interviews and conversations with researchers in the laboratory of neuroscientist Larry Young at Yerkes Primate Research Center, she raised questions about sex and gender frequently. They were met, generally speaking, with enthusiasm and understanding. They know better than the average non-women's studies major that sex is a fragile concept, its meaning contingent upon group consensus, sometimes as local as the lab. They know that genes are not determinants of gendered behavior. Willey experienced no great rift in our understandings of the social world and was able to explain her own academic interest in monogamy with relative ease. What she did encounter was a persistent gesture, for lack of a better word, to the results. Interviewees acknowledged that experiments are "quick and dirty" approximations of "real" processes and that, still, they work. We want to ask why they work. In this section we look at the sex-stratified and gendered experiments of the lab, based on titles and abstracts from their research publications. In the next, we zoom in on how this gendered logic reproduces itself in these experiments through assumptions about sex hormones.

Based on a Pubmed search for articles out of Young's lab that had to do with either oxytocin or vasopressin, we found sixteen primary research articles published since 2001 (see table). By reading the methods sections, we found that in five articles (31 percent) all of the subjects were female, in nine (56 percent) all subjects were male, and in two articles (13 percent) there were both male and female subjects. Four out of five of the female-only studies include "female" as a qualifier in the title of the paper. None of the nine male-only studies include a qualifier. In fact, most of these do not even specify that they are male-only in their abstracts. Female is a marked category of difference while maleness stands in as neutral. Presumably, this means that the female-only studies are only relevant for female populations and the male-only studies are more generalizable. We argue that this only makes sense from a distance. When you consider the hormones they study and the meanings attached to them, sexual difference emerges as a pursuit of monogamy gene research. Almost all of the studies on oxytocin were conducted in female-only populations, and almost all of the studies on vasopressin were conducted in male-only populations. Why is this and (why) does it "work"?

Young's laboratory's publications on oxytocin and vasopressin, 2001–2009

Source/year	*Title*	*Sex of subjects*	*Hormone system studied*
Ross, Cole, et al. 2009	"Characterization of the oxytocin system regulating affiliative behavior in *female* prairie voles"	Female	Oxytocin
Ross, Freeman, et al. 2009	"Variation in oxytocin receptor density in the nucleus accumbens has differential effects on affiliative behaviors in monogamous and polygamous voles"	Female	Oxytocin
Olazabal and Young 2006a	"Oxytocin receptors in the nucleus accumbens facilitate 'spontaneous' maternal behavior in adult *female* prairie voles"	Female	Oxytocin
Olazabal and Young 2006b	"Species and individual differences in juvenile *female* alloparental care are associated with oxytocin receptor density in the striatum and the lateral septum"	Female	Oxytocin

(*continued*)

Young's laboratory's publications (*cont.*)

Source/year	*Title*	*Sex of subjects*	*Hormone system studied*
Bielsky, Hu, Ren et al. 2005	"The V1a vasopressin receptor is necessary and sufficient for normal social recognition: A gene replacement study"	Male	Vasopressin
Bielsky, Hu and Young 2005	"Sexual dimorphism in the vasopressin system: Lack of an altered behavioral phenotype in *female* V1a receptor knockout mice"	Female	Vasopressin
Hammock et al. 2005	"Association of vasopressin 1a receptor levels with a regulatory microsatellite and behavior"	Male	Vasopressin
Hammock and Young 2005	"Microsatellite instability generates diversity in brain and sociobehavioral traits"	Male	Vasopressin
Nair et al. 2005	"Central oxytocin, vasopressin, and corticotropin-releasing factor receptor densities in the basal forebrain predict isolation potentiated startle in rats"	Male	Oxytocin and Vasopressin
Bielsky et al. 2004	"Profound impairment in social recognition and reduction in anxiety-like behavior in vasopressin V1a receptor knockout mice"	Male	Vasopressin
Lim, Murphy, et al. 2004	"Ventral striatopallidal oxytocin and vasopressin V1a receptors in the monogamous prairie vole (*Microtus ochrogaster*)"	Both	Vasopressin
Lim, Wang, et al. 2004	Enhanced partner preference in a promiscuous species by manipulating the expression of a single gene"	Male	Vasopressin
Lim and Young 2004	"Vasopressin-dependent neural circuits underlying pair bond formation in the monogamous prairie vole"	Male	Vasopressin
Phelps and Young 2003	"Extraordinary diversity in vasopressin (V1a) receptor distributions among wild prairie voles (*Microtus ochrogaster*): Patterns of variation and covariation"	Both	Vasopressin
Ferguson et al. 2001	"Oxytocin in the medial amygdala is essential for social recognition in the mouse"	Male	Oxytocin
Pitkow et al. 2001	"Facilitation of affiliation and pair-bond formation by vasopressin receptor gene transfer into the ventral forebrain of a monogamous vole"	Male	Vasopressin

Is Oxytocin from Venus?

According to researchers, the release of certain hormones during mating, between a male and female, causes the animals to bond—if they are monogamous. In females, the hormone said to facilitate pair bonding is oxytocin, which is also widely understood to control maternal care behaviors. Pair bonding in males is said to be facilitated by the hormone vasopressin. Vasopressin is not linked specifically to bonding or caregiving behaviors, but rather is said to control "species-specific" behaviors in males. As neuroscientist Larry Young explained in an interview, vasopressin makes hamsters territorial; it makes prairie voles bond with a female and want to protect her and their young. This research agenda reflects a story about monogamy that is strangely familiar—females become intensely attached to whoever they have sexual intercourse with and males do not (Dallos and Dallos 1997, 139–141). When males settle down, they are motivated by feelings of possessiveness. We offer analyses of two sets of assumptions informing this gendered story: (1) definitions of sex (the act) that serve as the rationale for oxytocin's centrality to pair bonding and (2) the notion of "the maternal brain." These assumptions are so deep and so persistent that evidence disrupting them tends to be ignored or compartmentalized, such that research on these "monogamy" hormones proceeds with veracity along sexed lines.

Heather Ross and Larry Young's account of how oxytocin became a prime candidate for pair-bonding research is telling: "Because of the role that [oxytocin] plays in mother-infant attachment and since mating results in vaginocervical stimulation, which is known to release oxytocin in the brain, oxytocin was a prime candidate for regulating the formation of the pair bond" (2009, 538). Two foundational assumptions are operating here: (1) mating is somehow causally linked to pair-bond formation and (2) mother-infant bonding is either similar to or linked with pair bonding. We want to question the obviousness of these implicit claims.

"Mating" causes a pair bond to form, and mating is coded as vaginal intercourse. For females "vaginocervical stimulation" is the behavioral mechanism that starts this process. Although it is not discussed to nearly the extent the cervix is, it would seem that orgasm is what triggers this neurochemical process for males. The existence of a potentially analogous phenomenon in females leads Young (2009), in a recent essay in the opinions section of *Nature*, to laud the evolutionary wonder of cervical stimulation as part of sexual intimacy.

The lack of widespread consensus about the joys of cervical exams or childbirth have led feminists to question evolutionary explanations of female sexuality that link it to, or rather explain it in terms of, reproductive sex.[3] Elisabeth Lloyd (1993) calls this "the orgasm-intercourse discrepancy." Specifically, she interrogates ideas about orgasm as an evolutionary adaptation that rewards female primates for having frequent sex with their male mates. The linking of orgasm to intercourse, she argues, is a fairly ludicrous yet totally unquestioned assumption underlying the formulation of questions and experiments across a broad range of research on female sexuality. A version of this "discrepancy" seems to be foundational to our understanding of the naturalness of female monogamy here.

In fact, a great many things trigger the release of oxytocin and could be understood to facilitate the formation of bonds between animals. Small amounts of oxytocin are released over time when the animals spend long periods of time together; nipple stimulation, touching, and "grooming" cause oxytocin to be released; and hormones released during exercise may also bind with these same receptors. [. . .]

Oxytocin continues to be understood as the cause of universal cross-species maternal behavior, even as conflicting evidence emerges within the lab. In 2001, Thomas Insel and colleagues reviewed evidence that surprised researchers in an article titled "Oxytocin: Who Needs It?" The authors concluded "not mice" based on a study that showed that mice with a null mutation for oxytocin "exhibited full maternal and reproductive behaviors" (Insel, Gingrich, and Young 2001, 59). They went on to suggest that primates may not need oxytocin for these behaviors either. These data, even as they are cited in future research papers, become inassimilable in the retelling of the story of oxytocin's role in maternal care: the narrative is consistently universalized and persistently cited as evidence of the efficacy of putting funding dollars into research on the hormone.

Despite highly sex-specific research on oxytocin, the hormone and receptors are found in both males and females. It has been reported that there are no sex differences in the distribution of oxytocin receptors in the brains of voles (Lim, Murphy, and Young 2004). The conclusion they draw is that some other part of the oxytocin system must be sexually dimorphic. So we begin with two assumptions: that males and females are different and that oxytocin is a female/maternal hormone. Based on these pre-theoretical truths, the search for sex differences and oxytocin's role in manifesting them is a central and ongoing tenet of vole research on monogamy.

A Persistent Narrative

[. . .] Is it possible that vasopressin and oxytocin are not so gendered as we imagine? What could we learn to know about human attachment, about "love," if we could imagine it outside of historically situated ideas about gender and heterosexuality? If the naturalness of maternal care, the link between love and reproduction, and the primacy of sexual difference were questions, not the stable ground from which we do the asking, what would we want to know?

Notes

1 The video can be viewed at www.youtube.com/watch?v=Oh8x9KDkYTc (accessed November 21, 2010).
2 See also Irigaray, 1985 (especially page 15) for another approach to deconstructing the logic of sperm and egg as active and passive.
3 See also Koedt, 1970.

Works Cited

Andrea and Alicia. 2008. "Oxytocin, Vasopressin, and a Tale of Two Voles." *The New View on Sex*. http://thenewviewonsex.blogspot.com/2008/04/oxytocin-vasopressin-and-tale-of-two.html.
Arrizón, Alicia. 2008. "Latina Subjectivity, Sexuality and Sensuality." *Women & Performance: A Journal of Feminist Theory* 18 (3): 189–198.
Ballon, M.S. 2005. "To Have and to Vole." June 18.
Dallos, Sally, and Rudi Dallos. 1997. *Couples, Sex, and Power: The Politics of Desire*. Philadelphia: Open University Press.
Emens, Elizabeth F. 2004. "Monogamy's Law: Compulsory Monogamy and Polyamorous Existence." *U of Chicago, Public Law Working Paper* (58): 277.
Hammock, Elizabeth AD, and Larry J Young. 2005. "Microsatellite Instability Generates Diversity in Brain and Sociobehavioral Traits." *Science* 308 (5728): 1630–1634.

Hantzis, D. M., and M. V. Lehr. 1994. "Whose Desire? Lesbian (Non) Sexuality and TV's Perpetuation of Hetero/Sexism." In *Queer Words, Queer Images: Communication and the Construction of Homosexuality*, edited by R. Jeffrey Ringer, 107–121. New York: NYU Press.
Hill Collins, Patricia. 1990. *Black Feminist Thought: Knowledge, Consciousness, and the Politics of Empowerment*. Boston: Unwin Hyman.
———. 2004. *Black Sexual Politics: African Americans, Gender, and the New Racism*. New York: Routledge.
Hubbard, Ruth. 1990. *The Politics of Women's Biology*. New Brunswick, NJ: Rutgers University Press.
Hubbard, Ruth, and Elijah Wald. 1999. *Exploding the Gene Myth: How Genetic Information is Produced and Manipulated by Scientists, Physicians, Employers, Insurance Companies, Educators, and Law Enforcers*. Boston: Beacon Press.
Insel, Thomas R, Brenden S Gingrich, and Larry J Young. 2001. "Oxytocin: Who Needs It?" *Progress in Brain Research* 133: 59–66.
Irigaray, Luce. 1985. *Speculum of the Other Woman*. Ithaca, NY: Cornell University Press.
Jacquet, L. 2005. *March of the Penguins*. Warner Independent Pictures.
Johnston, I. 2005. "How Geneticists Put the Romance Back into Mating." *The Scotcsman*, July 30.
Koedt, Anne. 1970. "The Myth of the Vaginal Orgasm." *Radical Feminism: A Documentary Reader*: 371–7.
Lewontin, Richard C. 1991. *Biology as Ideology: The Doctrine of DNA*. 1st U.S. ed. Toronto: House of Anansi.
Lim, Miranda M, Anne Z Murphy, and Larry J Young. 2004. "Ventral Striatopallidal Oxytocin and Vasopressin V1a Receptors in the Monogamous Prairie Vole (*Microtus ochrogaster*)." *Journal of Comparative Neurology* 468 (4): 555–570.
Lim, MM, EAD Hammock, and LJ Young. 2004. "The role of vasopressin in the genetic and neural regulation of monogamy." *Journal of Neuroendocrinology* 16 (4):325–332.
Lloyd, Elisabeth A. 1993. "Pre-theoretical Assumptions in Evolutionary Explanations of Female Sexuality." *Philosophical Studies* 69 (2):139–153.
Martin, Emily. 1991. "The Egg and the Sperm: How Science has Constructed a Romance Based on Stereotypical Male-Female Roles." *Signs* 16 (3):485–501.
Murray, A. 1995. "Forsaking All Others: A Bifeminist Discussion of Compulsory Monogamy." In *Bisexual Politics: Theories, Queries, and Visions*, edited by Naomi Tucker, xxviii, 358 p. [[AU: check page range]] New York: Haworth Press.
Ordover, Nancy. 2003. *American Eugenics: Race, Queer Anatomy, and the Science of Nationalism*. Minneapolis: University of Minnesota Press.
Phelps, Steven M, and Larry J Young. 2003. "Extraordinary Diversity in Vasopressin (V1a) Receptor Distributions Among Wild Prairie Voles (*Microtus*

ochrogaster): Patterns of Variation and Covariation." *Journal of Comparative Neurology* 466 (4): 564–576.
Rader, Karen A. 1995. "Making Mice: Standardizing Animals for American Biomedical Research." Indiana University.
Rich, Adrienne. 1980. "Compulsory Heterosexuality and Lesbian Existence." *Signs* 5 (4): 631–660.
Rosa, B. 1994. "Anti-monogamy: A Radical Challenge to Compulsory Heterosexuality." In *Stirring It: Challenges for Feminism*, edited by Gabriele Griffin, 107–120. London: Taylor & Francis.
Ross, Heather E, and Larry J Young. 2009. "Oxytocin and the Neural Mechanisms Regulating Social Cognition and Affiliative Behavior." *Frontiers in Neuroendocrinology* 30 (4): 534–547.
Sample, I. 2005. "Love is a Drug for Prairie Voles to Score." *Guardian*, December 5.
Snider, C. J. 2002. "What's Love Got to Do with It? (Cartoon)." *Science and Spirit Magazine*.
Young Lab. "Vole Genomics Initiative." Accessed September 30. http://research.yerkes.emory.edu/Young/volegenome.html.
Young, Larry J. 2009. "Being Human: Love: Neuroscience Reveals All." *Nature* 457 (7226): 148–148.

Natural Convers(at)ions

Or, What if Culture Was Really Nature All Along?

VICKI KIRBY

The Linguistic Turn—Culture Takes Precedence

The "linguistic turn" in postmodern and poststructural criticism has had a major impact on the landscape of the humanities and social sciences and the way we conceive and communicate our various concerns. Words such as "text," "writing," "inscription," "discourse," "language," "code," "representation," and so on are now part of the vernacular in critical discussion. Indeed, over the years the textualizing of objects and methodologies has generated new interdisciplinary formations across the academy and transformed the content, approach, and even the justifications for research. On the political front we have seen similar shifts in the practices, modes of argumentation, and even the alliances and strategies that once identified particular social movements and struggles for equity. And all this because the material self-evidence of initial conditions or first causes, those stable analytical parameters that allow us to identify a problem and then debate what needs to be done to correct it, has suffered a significant assault. Although in a very real sense political contestation has always debated first principles, once the substantive difference between nature and culture, or temporal priority and causal directionality is disestablished, we enter a very different zone of political possibility.

Abridged form of Vicky Kirby, "Natural Convers(at)ions: Or, What If Culture Was Really Nature all Along?" in Stacy Alaimo and Susan Hekman (eds), *Material Feminisms*, Indiana University Press, 2008, 214–237. Reprinted with permission.

The following meditation will revisit what has surely been a truism for cultural criticism, namely, the need to interrogate the nature/culture division and the entire conceptual apparatus that rests upon it. [. . .] This analysis will draw on one of the genetic markers of a certain style of feminist and cultural criticism, namely, the critique of Cartesian thought and the political inflections that pivot around its binary logic. Theorists of gender, sexuality, and race, for example, have found that Nature/the body is routinely conflated with woman, the feminine, the primordial, with unruly passion and "the dark continent"—all signs of a primitive deficiency that requires a more rational and evolved presence (the masculine/whiteness/heterosexuality/culture and civilization) to control and direct its unruly potential. The value of this work is not in dispute here; indeed, in a very real sense this paper will try to extend the more intricate and productive aspects of its insights. Nevertheless, the immediate task is to understand why, on closer inspection, the strategies for overturning the automatic denigration of Nature and the battery of devaluations associated with it have remained wedded to its repetition.

Since this is a big claim, it is best approached in small steps, steps that will retrace our commitments to some foundational building blocks. Let's begin with the problem of binary oppositions in order to understand why this logic might enable Cartesianism as well as the arguments in cultural criticism that strive to overturn it. For example, it is somewhat routine within critical discourse to diagnose binary oppositions as if they are pathological symptoms: conceptual errors that are enduring, insidious, and whose effects can normalize political inequity. However, if the remedial treatment for such symptoms is to replace these binary errors with nonbinary correctives, then surely we are caught in something of a quandary. In other words, if every maneuver to escape binary logic effectively reinstates it in a more subtle way, then perhaps we need a more careful examination of what we are actually dealing with.

To take just one facet of the binarity riddle, we might consider whether the difference that renders entities distinct and autonomous is a true reflection of their actual independence and separateness. This seems like a straightforward question, yet one of the insights in semiology is that when we identify something and attribute it with its very own meaning and properties, we arrive at this determination through an entangled knot of associations. The co-responding resonances that animate language and perception actually determine (some might say produce) particularity, and this is why certain poststructural accounts of identity formation argue that context, an external difference, is also constitutively and operationally interior to the identity it seems to surround. However, if we commit to the notion of difference as an internal ingredient in identity formation rather than an external "in-between" identity, should we then conclude that different identities must be inseparable rather than autonomous? The real curiosity appears at this juncture: we are now unable to hold on to the difference between separability and inseparability, between one identity and another, because we have just conceded that the effective difference within all binary oppositions, including this one, is profoundly compromised.[1]

As discussions of the Cartesian mind/body (nature/culture) division so often illustrate, the more counterintuitive and potentially productive dimensions of the binary puzzle that query the very makeup of the categories can often disappear in the diagnostics of critique: [. . .] instead of acknowledging that the very stuff of the body and the processes that purportedly separate thought from carnality are now something of a mystery, the essence of these "components" and their connections can be taken for granted. There is little risk in most contemporary criticism, for example, of attributing agency and intelligent inventiveness (culture) to the capacities of flesh and matter (nature). In sum, nature is deemed to be thoughtless, and political interventions into Cartesian logic are much more likely to preserve this assumption by expanding the category "culture" to include whatever it is defined against. If the myriad manifestations of nature are actually mediations or re-presentations, that is, second order signs of cultural invention, then nature, as such, is absent.

Although these analytical maneuvers represent crucial points of entry into the more fascinating implications of this problematic, it becomes clear that both Cartesianism and its critique are entirely committed to the difference between nature and culture, presence and absence, and matter and form. Arguing that we remain indebted to the materiality of the body, that we are always attached to it and never independent of it, that both women and men are equally corporeal, or that none of us can *properly* be identified with nature's primordial insufficiency if this determination is a political (cultural) one, doesn't in any way dislodge the premise of

Cartesianism. In all of these arguments it goes without saying that nature/the body/materiality preexists culture/intellect/abstraction, and furthermore, that the thinking self is not an articulation of matter's intentions. Given this, what we will need to keep at the forefront of this meditation is whether the conventional sense of difference as something that divides identities from each other—materiality *from* abstraction—or similarly, something that joins materiality *to* abstraction (because we still assume in this case that two different things are connected), can adequately acknowledge the riddle of identity.[2]

But let's return to the nature/culture division to consider how this particular example of identity that presumes opposition is commonly explained in cultural and feminist theoretical writings. In the main, it is now axiomatic to eschew naturalizing arguments for several reasons. First, and perhaps most important, they are regarded as inherently conservative. Compared with the cacophony of cultural explanations that exemplify contestation, movement, and change, it follows that natural determinations will seem like a prescriptive return to something from the past, something undeniable and immutable. In the former case, when we explain our thoughts and actions as cultural products and effects, we are also emphasizing that we are active agents in our political destinies. By embracing the notion of natural cause and determination, however, we run the risk of reducing what seems so special about the human condition to evolutionary happenstance, or nature's caprice. In a very real sense, then, it is the essential nature of human being that is at stake in these debates.

The assumption that the threat of nature can be put aside in some way has been justified theoretically by the linguistic turn itself, which promotes the belief that culture is an enclosed system of significations that affords us no immediate access to nature at all. According to this view and as already noted above, cultural webs of interpretation, which include linguistic and even perceptual frames of legibility, are intricately enmeshed and cross-referenced, and this raft of mediations stands between any direct experience or knowledge of nature's raw facticity. Consequently, the difference between cultural and natural facts is impossible to adjudicate, and this is why we inevitably confuse cultural constructions of nature with "Nature itself."

When Scientific Objects Turn Into Language

A clear illustration of the view that it is in the nature of culture to misrecognize culture as nature is evident in the following example. In an interview with Judith Butler, whose work is well known for its analytical commitment to cultural constructionism, I took the opportunity to ask if the organizing trope in her work, namely, language, discourse—textuality—had been too narrowly conceived. My question was inspired by medical and scientific research that claims to investigate the brute reality of material objects and processes: why does the essential nature of these scientific objects also appear to be textual? Within the sciences, the stuff of the body appears as codes, signs, signatures, language systems, and mathematical algorithms. In the cognitive sciences, for example, it seems that explanations of neural-net behavior, or how neurons learn new material (become different), parallels Ferdinand de Saussure's explication of the peculiar resonances of the language system (Wilson 1998, 189–98). Other useful comparisons have been made between the communicative structures of biological languages and the language theories of Charles Sanders Peirce (Hoffmeyer 1996). And Jacques Derrida acknowledged that the puzzle of language was just as evident in the biological sciences as it was in literature and philosophy.[3] Even the layperson is increasingly aware that biological information in general, from genetic structures to the translation capacities of our immune system, shares some workable comparison with natural languages. But what are these languages, these biological grammars that seem to be the communicative stuff of life?

Admittedly, we don't tend to think of signs as *substantively* or ontologically material. But what prevents us from doing so? With such considerations in mind, I directed the following question to Butler: "There is a serious suggestion that 'life itself' is creative encryption. Does your understanding of language and discourse extend to the workings of biological codes and their apparent intelligence?" (Breen et al. 2001, 13). On this last point, I was thinking of the code-cracking and encryption capacities of bacteria as they decipher the chemistry of antibiotic data and reinvent themselves accordingly. Aren't these language skills?

Butler's response is a form of admonition, a reminder that language is circumscribed, that its author and reader is human, and that the human endeavor to capture a world "out there" through cultural signs will always be a failed project. To this end, she warns:

> There are models according to which we might try to understand biology, and models by which we might try to understand how genes function. And in some cases the models are taken to be inherent to the phenomena that is *[sic]* being explained. Thus, Fox Keller has argued that certain computer models used to explain gene sequencing in the fruit fly have recently come to be accepted as intrinsic to the gene itself. I worry that a notion like "biological code," on the face of it, runs the risk of that sort of conflation. I am sure that encryption can be used as a metaphor or model by which to understand biological processes, especially cell reproduction, but do we then make the move to render what is useful as an explanatory model into the ontology of biology itself? This worries me, especially when it is mechanistic models which lay discursive claims on biological life. What if life exceeds the model? When does the discourse claim to become the very life it purports to explain? I am not sure it is possible to say "life itself" is creative encryption unless we make the mistake of thinking that the model is the ontology of life. Indeed, we might need to think first about the relation of any definition of life to life itself, and whether it must, by virtue of its very task, fail. (Breen et al. 2001, 13)

Butler is understandably vigilant about the seductive slide that conflates representations, models, and signs that substitute for material objects, with the objects themselves. In other words, although it is inevitable that we will misrecognize one in the other, Butler cautions against committing to the error. When dealing with scientific objects, the transparent self-evidence of reality is even more persuasive, but even here we are encouraged to remember that these objects are actually literary—textual, or encoded forms of language—and to this extent, if they can only emerge through cultural manufacture, then their reality and truth is attenuated, or even illusional.

Although this argument is certainly persuasive, especially against the sort of hard-edged empiricist and positivist scientific claims that give little consideration to the vagaries of interpretation, there are lingering problems nevertheless. If we contextualize Butler's intervention in terms of the political legacy of binarity mentioned earlier, she effectively challenges the devaluation of nature (the feminine, matter, the origin) by arguing that these significations are cultural ascriptions with no essential truth. If the economy of valuation can be analyzed, contested, and redistributed (because this is the operational definition of culture), then the question of nature is *entirely* displaced: put simply, it can have no frame of reference that isn't properly cultural. Indeed, even the concept/word "nature" is misleading because it evokes meanings, prejudices, and even perceptions that are learned and therefore inherently historical/cultural.

To accept that we are bound within the enclosure of culture is to commit to a raft of related assumptions, and although there is certainly some interpretive play in what we make of them, it might be helpful to register something of their broad outline here. The most important is the assertion that humanness is profoundly unnatural. The abstracting technology of language, intelligence, and creative invention is separated from the body of the material world, indeed, from the material body of human animality. Ironically, given the initial concern to question the separation of nature from culture within Cartesianism, the sense that human identity is somehow secured and enclosed against a more primordial and inhuman "outside" (which must include the subject's own corporeal being!) recuperates the Cartesian problematic, but this time without question. Given this, it is not surprising that cultural arguments that relentlessly interrogate the autonomy and integrity of identity formation fall mute when it comes to the question of how culture conceives and authenticates its own special properties and self-sufficiency. If we translate the separation of culture from nature into the mind/body split, it seems that the Cartesian subject can admit that s/he has a body (that *attaches* to the self), and yet s/he is somehow able to sustain the belief that *s/he is not this body.* This denial is necessary because to contest the latter and all its possible consequences would at least suggest that it might be in the nature of the biological body to argue, to reinvent, and rewrite itself—to cogitate.

Neither Descartes, nor any cultural critic who draws analytical purchase from some version of the linguistic turn, would deny that human identity incorporates two quite different systems of endeavor. Not many would dispute the presence of a biological reality that is quite different from culture and that we imperfectly try to comprehend. But surely, if we were without our skin and we could witness the body's otherwise invisible processes as we chat to each other, read a presentation aloud, type away at our computers, or negotiate an intense exchange with someone we care about, we might be forced to acknowledge that perhaps the meat of the body *is* thinking material. If it is in the nature of biology to be cultural—and clearly, what we mean by "cultural" is intelligent, capable of interpreting, analyzing, reflecting, and creatively reinventing—then what is this need to exclude such processes of interrogation from the ontology of life? The difference between ideality and matter, models and what they purportedly represent, or signs of life and life itself, is certainly difficult to separate here. However, it is important to emphasize that this confusing implication can't be corrected in the way that Butler attempts to do. Although her work underlines why there will always be confusion, she explains this blurring of object and interpretation as an inevitable mistake that derives from the human condition and the hermetic enclosure of the interpretive enterprise, or mind itself.

Entangling the Question of Language/System

Two different considerations arise at this juncture that assist in pushing the problematic forward instead of reiterating the normative frame of reference that is increasingly routine in cultural criticism. In passing, we might note that in a very different field of inquiry the implication between concepts (ideality) and things (materiality) in quantum theory is so profound that it undermines our understanding of their respective differences. Space re-forms as phenomena turn out to be specific and local as well as general and ubiquitous. And similarly, the temporal differences that separate past, present, and future appear to be synchronized when thought experiments can anticipate what will have already taken place: remarkably, the results of these experiments are retrospectively actualized and empirically verifiable.[4] Is the weirdness of this evidence rendered explicable because it reflects the epistemological intertextualities—the crisscrossings of metaphors and models whose cultural origins have little if anything to do with the actualities of the universe at large? Or is there a more worldly form of intertextual referencing in these scientific results that collapse concept (model) and thing, and disperse authorship, identity, and causality?

To underline the counterintuitive complexities in this question, we need to appreciate why the description of the thought experiment that can retrospectively anticipate and materialize what will already have taken place, can inadvertently ignore the mysterious sense of entanglement in this "event's" operational possibility. There is a temporal configuration in the above description's narrative order that preserves the logic of causal separation and the presumption that there are different moments in time, different places in space, and a very real difference between thought and material reality. To suggest that one affects the other in a way that renders them inseparable doesn't confound the nature of their difference (respective identities) so much as it emphasizes that these differences are joined, or connected in some way. In the first instance, we interpret "inseparability" to mean that human agency and intention produced a change in the nature of reality and that this is proof of some mysterious connection between them. The very same logic would allow us to reverse the direction of this causal explanation to suggest that some agential force in the universe directed humans to conduct an experiment whose results the world had already anticipated. However, neither of these explanations captures the space/time entanglement at work here, even though this last reversal begins to trouble the properties that we tend to attribute to these different identities (human and nonhuman) and the relational asymmetries that affirm the difference in their properties, capacities, and timings.

If we consider these implications more rigorously, then significance and substance, thought and matter, human agency and material objectivity, must be consubstantial. But what does this actually mean, and can we do anything interesting with such a wild assertion? Of course, the linguist Ferdinand de Saussure said something very similar about the consubstantiality of semiological entanglement, just as Judith Butler's more contemporary interpretation of his argument insists that signification matters and that ideation can realize. In other words, many of the most important interventions in cultural criticism that condense differences together seem to mimic such counterintuitive assertions. Our question is whether these insights only relate to the peculiar attributes of culture. To return to the question of scientific modeling, must we assume that these models are interpretive illusions produced by humans to mirror a world that can't be accessed? Because if they are mere illusions and the world is not present *in* them, then how can they possess the extraordinary capacity (as we see in the case of quantum relations) to anticipate verifiable outcomes whose pragmatic results heralded contemporary advances in computer and electronic technologies?

It seems that the little steps by which we retraced our way to certain foundational commitments about binarity and the nature of language very quickly turned into the most puzzling quandaries about the nature of life and the mysteries of the universe! And while the complexities of scientific theory surely exceed our disciplinary expertise, the discussion above has made the appeal to an "absolute outside" of anything, whether the discipline of physics, or theories of textual interpretation, for that matter, considerably more fuzzy. For this reason, and in the spirit of a more meditative style of inquiry, perhaps we can at least risk the suggestion that if the quantum conflation of *thesis* with/in *physis* has general purchase, then we should not read the most complex aspects of poststructuralism as pure *thesis*. The most counterintuitive arguments about the superposition of matter and ideation, concept and object; all of the close analytical criticism that discovers systems of referral and relationality *within* identity/the individual; the peculiar space/time condensations that we confront in Freud's notion of memory or *nachträglichkeit* (deferred action); or the "intra-actions"[5] of Derridean *différance* and its counterintuitive implications—need we assume that such insights are purely "cultural" because the world itself, in its enduring insistence, simply couldn't be that dynamically involved and alien to ordinary common sense?

But let's stick to something more straightforward that will test the conventional interpretation of cultural constructionism just as effectively by showing that a precritical understanding of reference as something self-evident in nature is inadequately countered by theories of the referent as a cultural artifact. The question is disarmingly simple, so simple that one has to wonder why such questions are so rarely asked within the disciplinary protocols of cultural criticism; namely, can the rampant culturalism that understands the mediations of language as a purely cultural technology, a technology that *cannot* have any substantial purchase because it remains enclosed against itself, explain how computational models, bio-grams of skin prints, blood evidence, genetic signatures, pollen chemistries, and insect life cycles (all data that present as languages)—how can this cacophony of differences possess *any* possibility of predictive reference? As we are well aware from forensic investigation techniques, data is indicative. From global networks of information that bring geology, biology, psychology, entomology, cryptography, and even the very personal street-smarts of a particular observing investigator into conversation and convergence, a referent is thrown up. The most obvious question that this intricate process raises is—how?

Quantum Implications and the Practice of Critique

The reluctance of postmodern styles of criticism to actively consider how scientific models of nature work at all (even when imperfectly, but certainly when we witness their extraordinary predictive accuracy) has led many to discount the productive energy in these theories without appreciating what they can actually offer. Bruno Latour, for example, a sociologist and historian of science, pours vitriol upon those "gloating" cultural constructivists whose smug self-enclosure attributes all agency and articulation to a brain in a vat (1999, 8). With considerable irritation, he rails against the idea that anyone could celebrate musing blindly about a world that can no longer be accessed from the confines of a linguistic prison house. To paraphrase his position, such arguments descend further and further into the same dark and spiraling curves of the same hell that is Cartesianism—"We have not moved an inch" (8).

[. . .] Unlike conventional postmodern approaches, Latour's approach doesn't emphasize the vagaries of subjectivism and the relative illusiveness of truth. Such a claim would reiterate Butler's position—that an essential and natural truth is veiled behind culture's misguided attempts to represent it. Instead, Latour effectively redefines "the social" in a more comprehensive way—as a confluence of forces and associations, a collective assembly of human *and* nonhuman interactions that together produce social facts with referential leverage. [. . .][6]

When we posit a natural object, a plant, for example, we don't assume that it is unified and undifferentiated: on the contrary, this one thing is internally divided from itself, a communicating network of cellular mediations and chemical parsings. It is a functioning laboratory, a technological apparatus whose intricate operations are finely elaborated—an "intermediate" node that communicates its ecological significance in a way that incorporates and blurs the outside with/in the inside. Given this, why is it so difficult to concede that nature already makes logical alignments that enable it to refer productively to itself, to organize itself so that it can be understood . . . by itself? If we turn to Latour's explanation of circulating reference as a mutual "construction" of human *and* nonhuman in the soil study (1999), it is clear that the soil offers itself as the material origin or object *for* study. Locked in at one end of a continuum, nature needs to be cultivated, cultured, and coaxed to reveal its secrets. Its lessons are educed by something that, inasmuch as it has the capacity to reveal and encode, cannot be a natural operation by definition. For Latour, then, nature is not itself a laboratory, an experimental, communicative enterprise.

I could conclude the argument by suggesting that nature doesn't require human literary skills to write its complexity into comprehensible format. But if I did this, I'd actually be reiterating the premise of Latour's position all over again by dividing "human" from "nonhuman": those "nonhumans" simply don't need us . . . don't be so pompous in assuming that they do! But perhaps there is a position that can affirm the human, *with* Latour, and even with a sociobiological twist that will address Latour's concerns about such "science fundamentalists" and allow their work to challenge us and not simply define our value against their foolishness.

My suggestion is to try a more counterintuitive gambit, namely, to generalize the assumed capacities of humanness in a way that makes us wonder about their content—after all, what do we really mean by agency, distributed or otherwise, or by intentionality and literacy? For example, why condemn the sociobiologist E. O. Wilson because he sees humanity reflected in the behavior of ants, in their animal husbandry and finessed horticultural skills, in the political complexities of their caste system, their slaving behaviors, the adaptive ministrations of their nursery regimes, their language and culture? When we explain this social complexity as an anthropomorphic projection whose comparison diminishes what is specific to human be-ing, we automatically secure the difference of our identity *against* the insect (nature) and reiterate that agrarian cultivation and animal husbandry (culture) first appeared with Neolithic peoples. We hang on to such assumptions by insisting that natural "smarts," clear evidence of engineering intelligence, social complexity, ciphering skills, and evolutionary innovation are just programs, the mere expression of instinctual behaviors. It is understandable why both Butler and Latour, for that matter, would reject the suggestion that human subjectivity, self-consciousness, and agency are "mere" programs. But what is a program if it can rewrite itself? Certainly not *pre*-scriptive? Surely, the point isn't to take away the complexity that culture seems to bring to nature but to radically reconceptualize nature "altogether."

The distributed agency of a "human nature" would "act, or communicate, at a distance." This quantum puzzle is actively embraced in "Circulating Reference," where Latour ponders how soil samples taken from the Brazilian savanna can maintain ontological constancy through the variety of instrumental translations, representations, and transformations they undergo—from soil, crumbling between the researchers' fingers, to its final recordings on many sheets of paper. Latour assures us that "here it is no longer a question of reduction [of the soil into words and graphs] but of transubstantiation" (1999, 54). Transubstantiation is a religious term, and yet one that could just as well be applied to quantum phenomena. It certainly evokes an abyssal crossing; however, this is not the gulf between nature and culture that Butler finds insurmountable, nor is it the gulf between nature and culture across which Latour discovers many bridges of cooperation. This radical disjunction/inseparability is comprehensive—a fault line that runs throughout all of human nature. It articulates the nonlocal within the local, nature within culture, and human within nonhuman. The superposition of these differences means that any identity is articulated with and by all others—consubstantiality and Latour's transubstantiation are one and the same. This is a comprehensive process, a process of comprehension, a material reality.

What happens if nature is neither lacking nor primordial, but rather a plenitude of possibilities, a cacophony of convers(at)ion? Indeed, what if it is that same force field of articulation, reinvention, and frisson that we are used to calling—"Culture"? Should feminism reject the conflation of "woman" with "Nature," or instead, take it as an opportunity to consider the question of origins and identity more rigorously?

Notes

1 For a detailed discussion of this puzzle in regard to the identity of the sign, see (Kirby 1997).
2 Jacques Derrida's early work on the logic of the supplement is especially pertinent here (Derrida 1984).
3 Although the point is made in passing in *of Grammatology* (1984, 9), Derrida specifically addresses this connection in a series of seminars on the Nobel

Prize winner François Jacob, who worked on the language of RNA. To date, the seminars remain unpublished.

4 Experiments undertaken by Alain Aspect and, more recently, Nicolus Gisin have confirmed that non-locality is a general property of the universe. Consequently, if any "event" in the universe is inseparable from another, any part inseparable from the whole, then the local is articulated through the universal and vice versa. This rather extraordinary suggestion compromises spatial divisions and temporal differences: the notion of individuated events *in* time or *in* space is imploded. For a helpful introduction to this field of inquiry, see (Nadeau and Kafatos 2001).

5 Just as Derrida conceived the neologism *différance* to complicate the meaning of difference, so Karen Barad has coined the term "intra-action" to evoke an involvement that is inadequately accommodated by the term "interaction" (2007).

6 *Editors' note:* For an in-depth discussion of Latour and a modified discussion of the concept of the quantum, see (Kirby 2011).

Works Cited

Barad, Karen Michelle. 2007. *Meeting the Universe Halfway: Quantum Physics and the Entanglement of Matter and Meaning*. Durham: Duke University Press.

Breen, Margaret Soenser, Warren J Blumenfeld, Susanna Baer, Robert Alan Brookey, Lynda Hall, Vicky Kirby, Diane Helene Miller, Robert Shail, and Natalie Wilson. 2001. "Introduction: 'There Is a Person Here': An Interview with Judith Butler." *International Journal of Sexuality and Gender Studies* 6 (1–2): 7–23.

Derrida, Jacques. 1984. *Of Grammatology*. Translated by Gayatri Chakravorty Spivak. Baltimore: Johns Hopkins University Press.

Hoffmeyer, Jesper. 1996. *Signs of Meaning in the Universe*. Bloomington: Indiana University Press.

Kirby, Vicki. 1997. *Telling Flesh: The Substance of the Corporeal*. New York: Routledge.

———. 2011. "Natural Convers(at)ions: Or, What if Culture Was Really Nature All Along?" In *Quantum Anthropologies: Life at Large*, xiv, 165 p. Durham: Duke University Press.

Latour, Bruno. 1999. "Circulating Reference: Sampling the Soil in the Amazon Forest." In *Pandora's Hope: Essays on the Reality of Science Studies*, 24–79. Cambridge, MA: Harvard University Press.

Nadeau, Robert, and Minas C. Kafatos. 2001. *The Non-local Universe the New Physics and Matters of the Mind*. Oxford: Oxford University Press.

Wilson, Elizabeth A. 1998. *Neural Geographies: Feminism and the Microstructure of Cognition*. New York: Routledge.

"The Bio-Technological Impact" *and* "Abstract Sex"

LUCIANA PARISI

The Bio-technological Impact

> Artifice is fully part of nature.
>
> DELEUZE, *BERGSONISM*

In 1985, Donna Haraway's Cyborg Manifesto highlighted the new mutations of the body-sex in bio-informatic capitalism. For Haraway, the convergence of bodies and technologies marked the emergence of the new metamorphic world of the cyborg, a hybrid blending of animal, human and machine parts. No longer embedded in the nuclear Oedipal family (the natural ties with the mother and the father), the cyborg was, for Haraway, the offspring of the post-gender world of genetic engineering where biological or natural sex no longer determines the cultural and social roles of gender (1985). Cybernetic communication and reproduction enable the prosthetic manipulation of the physical bonds of gender stretching the limits of Mother Nature. Artificial sex permits the unprecedented transformation of our gender identity, the construction and reconstruction of sexual forms and functions of reproduction.

The post-gender world of the cyborg brings to the extreme postmodern claims about the end of certitudes where biological destiny is threatened by the saturating proliferation of technologies of communication and reproduction in our daily life. As opposed to the postmodern nostalgia for a lost world of stable boundaries between nature and culture, the cyborg embraced the challenge of bio-informatic technologies affirming that our assumptions about nature are the results of intricate cultural constructions articulated by specific technoscientific discourses.[1] The equation between sexual identity and sexual reproduction at the core of our understanding of human sex is nothing natural. Quite the contrary, it is embedded in the historical and cultural roots of the Western metaphysical tradition of essentialism. Far from reflecting a given unquestionable truth, the cyborg revealed that the natural essence of a body rather derives from specific historical and cultural constructions (or representations) of nature establishing a natural association between feminine sex and sexual reproduction. Rather than being determined by sexual identity and sexual reproduction, the artificial world of the cyborg announces the new historical and cultural conditions of the posthuman body no longer able to find shelter in the natural world (Hayles 1999). For the postgender world of the cyborg, there is nothing natural about the human body, sex, and reproduction.

Haraway's seminal text has strongly influenced debates about the impact of bio-technologies on the body, sex and femininity. In particular, in the last ten years, debates about the convergence between biology and technology have problematized the new tension between natural and artificial sex, the disappearance of biological difference and the celebration of artificial disembodiment (Balsamo 1996; Stone 1991; Springer 1996; Turkle 1995; Squires 1996, 194–216; Braidotti 1994). It has been argued that the postgender world of the cyborg risks dissolving the biological differences of the body, the ties with the corporeal world of sex, celebrating the disembodied model of male pleasure (the independence from matter celebrated by the closed economy of charge and discharge). While liberating feminine desire from biological identity, the cyborg also deliberates the ultimate detachment of the mind from the body, the triumph of mental projections over material constraints (Flanagan and Booth 2002; Hayles 1999; Plant 2000, 1998, 1997).

These controversial debates about the implications of information technologies for sexual reproduction tend to perpetuate a critical impasse between biological essentialism and discursive constructivism. Claims about the return to material embodiments (biological differences) are opposed to the emergence of a post-gender world of cybersex where variable meanings and shifting discourses enable us to perform our gender identity beyond biological anatomy. In this framework, gender no longer depends on sex—the form of sexual organs and the function of sexual reproduction—rather it is sex that depends on the constructions of gender, the signifying signs that constantly change the nature of sex.[2] In recent years, the idea that you can perform your own gender by changing your sexual identity has strongly clashed with the feminist argument of maintaining biological ties among women in order to resist the accelerating disembodiment of difference in cybernetic capitalism.

Yet this critical impasse is nothing new. The constitution of binary oppositions between what is given (the natural or biological realm) and what is constructed (the cultural or technological world) is entangled with the traditional Western model of representation. As often argued, the model of representation does not entail the exact reflection of reality or truth, but is more crucially used to refer to a system of organization of signs where structures of meaning arrange gestural, perceptual, cognitive, cultural and technological signs through the hierarchies of the signifier (Guattari 1984, 73–81). The model of representation reduces all differences—biological, physical, social, economical, technical—to the universal order of linguistic signification constituted by binary oppositions where one term negates the existence of the other. The binary opposition between embodiment and disembodiment is caught up in the binary logic of representation that disseminates the dichotomy between materiality and immateriality, the separation of the inert body from the intelligent mind. Embedded in the Platonic and Cartesian metaphysics of essence, the logic of representation subjects the body, matter and nature to the transcendent order of the mind,[3] suppressing the network of relations between nature and culture, sex and gender, biology and technology, rapidly transforming the way we conceive and perceive the body-sex.

Neither the politics of embodiment nor disembodiment provides alternative conceptual tools to analyze the recent bio-informatic mutations of posthuman sex. This critical impasse is embedded in a specific conception of the body where a set of pre-established possibilities determines what a body is and can do. These possibilities are defined by the analogy between biological forms (species, sex, skin color and size) and functions (sexual reproduction, organic development and organic death) that shape our understanding of nature and matter through principles of identity (fixity and stability). This analogy creates a direct resemblance between body and mind, sex and gender, skin and race where biological destiny determines the hierarchical organization of social categories. Feminists and cyberfeminists have strongly criticized this biological sameness that constitutes the patriarchal model of representation whereby the body is mastered by the mind. Nevertheless, recent debates about cybersex or artificial sex have failed to provide an alternative understanding of the mind-body binarism reiterating the opposition between biological presence and discursive absence of the body.

The liberation from the mind-body dualism through the displacement of signifiers from fixed meanings (the signifier sex from the signified gender) appears to re-entrap the body in a pre-established set of possibilities determined by linguistic signification. The post-gender feminist attempt at untangling feminine desire from nature, through the floating of free signifiers of sex in the new cyberspace of information, problematically reiterates the mind-body dualism by associating the body with a fixed and stable nature where matter is inert. In a sense, post-gender feminism risks confusing the biology of the body with the materiality of a body where the conception of nature and matter is determined by and reduced to biological discourses or universal systems of signification. The continuous displacement of the signifier "sex" does not succeed in detaching feminine desire from fixed nature as it fails to challenge the fundamental problematic of the body, biological identity, the imperative of sexual procreation and ultimately the metaphysical conception of matter.

The bio-technological mutations of human sex and reproduction expose new implications for the separation of feminine desire from biological destiny requiring an altogether different conception of the body in order to challenge traditional assumptions (pre-established possibilities) about what we take a body to be and to do. Expanding upon the feminist politics of desire, abstract sex brings into question the pre-established biological possibilities of a body by highlighting the non-linear dynamics and the unpredictable potential of transformation of matter. Drawing on an alternative conception of nature, abstract sex embraces the Spinozist hypothesis about the indeterminate power (or abstract potential) of a body suggesting that "we do not yet know what a body can do." This hypothesis challenges the analogy between biological forms and functions (the pre-established biological possibilities of a body) pointing to the capacities of variation of a body in relation to the continual mutations of nature. Moving beyond the critical blockage between biological essentialism (embodiment) and discursive constructivism (disembodiment), abstract sex proposes a third route to widen the critical spectrum of our conception of the body-sex.

By proposing to re-wind the processes of evolution of the body and sex, abstract sex starts from the molecular dynamics of the organization of matter to investigate the connection between genetic engineering and artificial nature, bacterial sex and feminine desire that define the notion of a virtual body-sex. This notion is not to be confused with the immaterial body-sex as defined by the debates about the embodiment (materiality) and disembodiment (immateriality). The notion of the virtual body-sex primarily implies that a body is more than a biological or organic whole, more than a self-sufficient closed system delimited by predetermined possibilities. The virtual body-sex exposes the wider layers of organization of a body that include the non-linear relations between the micro level of bacterial cells and viruses and the macro levels of socio-cultural and economic systems. The collision of these layers defines the indeterminate potential of a body to mutate across different organizations of sex and reproduction producing a series of micro links between biology and culture, physics and economics, desire and technologies. The networked coexistence of these levels contributes to construct a new metaphysical conception of the body-sex that radically diverges from the binary logic of the economy of representation.

Abstract sex suggests that bio-technologies do not reiterate new or old dichotomies. Abstract sex displays the intensive connections between different levels of organization of a body-sex, where nature no longer functions as the source of culture, and sex of gender. The intensive concatenation between nature and culture entails reversibility in the ways in which nature affects and is affected by culture. This mutual relation points to an alternative understanding of sex and gender that no longer depends upon the primacy of identity and its mind-body binarism, but lays out the reversal relations between parallel modes of being and becoming of a body. Sex is neither constructed as the prediscursive or as the product of technoscientific discourses. Primarily sex is an event: the actualization of modes of communication and reproduction of information that unleashes an indeterminate capacity to affect all levels of organization of a body—biological, cultural, economical and technological. Sex is a mode—a modification or intensive extension of matter—that is analogous neither with sexual reproduction nor with sexual organs. Sex expands on all levels of material order, from the inorganic to the organic, from the biological to the cultural, from the social to the technological, economic and political. Far from determining identity, sex is an envelope that folds and unfolds the most indifferent elements, substances, forms and functions of connection and transmission. In this sense, sex—biological sex—is not the physical mark of gender. Rather, gender is a parallel dimension of sex entailing a network of variations of bodies that challenge the dualism between the natural and the cultural. Adopting Spinoza's ethics or ethology of the body, it can be argued that sex and gender are two attributes of the same substance, extension and thought, mutually composing the power—*conatus*—of a mutant body (Deleuze 1988b; Gatens 1996, 162–87). This conception of sex diverges from the critical impasse in cyberculture between essentialism and constructivism and its negative principles of identity.

From this standpoint, the bio-technological disentanglement of sex from sexual reproduction does not imply the ultimate triumph of the patriarchal model of pleasure, a longing for disembodiment and self-satisfaction. This disentanglement suggests an intensification of desire in molecular relations such as those between a virus and a human, an animal cell and a microchip. As opposed to the dominant model of pleasure defined by autoeroticism (the channeling of flows towards climax or the accumulation and release of energy), abstract sex points to a desire that is not animated or driven by predetermined goals. [. . .] Desire is autonomous from the subject and the object as it primarily entails a non-discharging distribution of energy, a ceaseless flowing that links together the most indifferent of bodies, particles, forces, and signs. In this sense, the cybernetic mutations of sex expose a continuum between the cellular levels of sex (bacterial sex), the emergence of human sex (heterosexual mating) and the expansion of bio-technological sex (cloning) entailing a new conception of the body. This conception highlights an alternative metaphysics of matter-nature that enfolds the multiple layers of composition of a body and sex, defining their potential capacity to differentiate.

Abstract sex points to the non-linear coexistence of the biophysical (the cellular level of the body-sex defined by bacteria, viruses, mitochondrial organelles, eukaryotic cells); the biocultural (the anthropomorphic level of the human body-sex defined by psychoanalysis, thermodynamics, evolutionary biology and anatomy in industrial capitalism); and the biodigital (the engineering level of the body-sex defined by information science and technologies such as in vitro fertilization, mammal and embryo cloning, transgenic manipulation and the human genome in cybernetic capitalism) layers of the virtual body-sex. This complex composition of the body-sex exposes the continual and unpredictable mixtures of elements stemming from different layers that indicate the indeterminate potential of a body-sex to mutate. In particular, the biotechnological engineering of the body, the genetic design of life accelerates the recombination of different elements and the mutations of the body-sex by disclosing a new set of urgent questions about the relation between feminine desire and nature.

The rapid innovations of cloning techniques seem to announce the ultimate achievement of Man over Nature, the ultimate power of Man to design Man. Yet, what might seem the final act of mastering nature by patriarchal humanism exposes in fact much more controversial implications. As pro-cloning and anti-cloning groups often point out, the genetic designing of life, involving the non-linear transfer of information between different bodies (animal, humans and machines), implies an acceleration of evolutionary mutations whose results are not yet known. The acclaimed final control of man over nature rather suggests the loss of human control on the unpredictable mutations of the body. The recent proliferation of mutant bodies radically brings into question the conception of nature where the acceleration of cloning, in the form of bacterial sex, suggests that artifice has always been part of nature. This rapid unfolding of artificial nature opens up new problematic questions in relation to bio-technological mutations of human sex and reproduction. If cloning has always been part of nature, as bacterial sex demonstrates, then isn't it natural to clone humans? Are the new bio-technologies of the body already part of nature? What are the implications of this newly defined artificial or engineering nature in relation to feminine desire? [. . .] By analyzing the implications of the bio-technological mutations of a body, abstract sex maps a wider critical route to relate the (cyber)feminist politics of desire with the artificiality of nature.

Abstract Sex

> The logos is a huge Animal whose parts unite in a whole and are unified under a principle or a leading idea; but the pathos is a vegetal realm consisting of cellular elements that communicate only indirectly, only marginally, so that no totalization, no unification, can unite this world of ultimate fragments. It is the schizoid universe of closed vessels, of cellular regions, where contiguity itself is a distance: the world of sex.
>
> DELEUZE, *PROUST AND SIGNS*

> With machines the question is one of connection or non-connection, without conditions, without any need to render an account to a third party. It is from that that the surplus value of encoding originates. The situation is like that of a bumble-bee which, by being there, became part of the genetic chain of the orchid. The specific event passes directly into the chain of encoding until another machinic event links up with a different temporalization, a different conjunction.
>
> GUATTARI, *MOLECULAR REVOLUTION: PSYCHIATRY AND POLITICS*

The mutations of a body are not predetermined by a given ideal or an infrastructure defining the realm of biological possibilities of a body. On the contrary, these mutations designate the abstract or virtual operations of matter. As Deleuze and Guattari argue, inspired by Henri Bergson, the virtual is not to be confused with the realm of the possible. The possible, in fact, is often the reflected image of an already determined reality contained in a closed set of choices. Possibilities do not have a reality, as their reality is already determined. Instead of denoting a possible reality, the virtual *is* reality in terms of strength or potential that tends towards actualization or emergence. Thus, the virtual does not have to become real. It is already real. It has to become actual. The actual does not derive from another actual, but implies the emergence of new compositions, a becoming that responds to (acts back on) the virtual rather than being analogous to it. Hence, virtuality and actuality do not coincide. They are two asymmetrical yet coexistent planes of difference that constitute the potential of a body to become different, to mutate beyond principles of analogy and resemblance.[4] Far from opposing matter to immateriality, abstract sex points to the potential mutations of a body that are not defined by a transcendent substance but by the incorporeal (abstract) transformations of matter (Deleuze 1990, 4–11, 13–21, 23, 35, 12–22, 67–73, 52–57; Foucault 1977, 165–99; Deleuze and Guattari 1988, 80–83, 85–88, 107–9).

Abstract matter is not substance. In the Cartesian tradition, substance corresponds to the non-extended God separated from the physical world of nature. The Cartesian split between the mind and the body originates from the separation of the cosmos from matter, of the transcendent God (the power of the soul-mind) from nature (the power of the physical body). In this framework, what we see in nature was created by a non-physical God, a superior entity that has the power to create and destroy the natural world. Contrary to Descartes's ideal soul, Baruch Spinoza's concept of substance demonstrates that nature is not separated from the cosmos. The body originates in God as God corresponds to an intensive and extensive substance. God does not create matter, but *is* matter able to manifest itself through the ceaseless mutation of bodies and things in nature. As explained later in this chapter, far from starting from the unity of the One, Spinoza points to the parallel multiplicities of being and becoming, the continual relations between the cosmos and nature, intensity and extension, mind and body that define the primacy of potential over possible matter. Abstract matter questions the philosophical tradition that separates the corporeal from the incorporeal, nature from culture, the organic from the technical. It exposes the potential relations of change between the virtual body and actual body, the symbiotic merging of non-identical powers (the continual power or potential between substance and modes) unfolding the unpredictable mutations of a body.

From this standpoint, abstract (mutating) matter is *machinic* as it entails the heterogeneous composition or merging of different bodies of production. This machinic process has nothing to do with the celebration of technological determinism where technical machines are opposed to the organic body (technology versus biology). Drawing on Deleuze and Guattari, a machine is above all defined by a mixture of biological, technical, social, economic and desiring elements that compose and decompose a body at certain speeds and according to given gradients. These mixtures are productive concatenations or *machinic assemblages* constituting for example the biocultural organization of the body (the disciplinary order of human sex established by the virtual links between psychoanalysis, anatomy, evolutionary biology and thermodynamics) that unleash a potential transformation of all the elements participating in the composition (the transformation of evolutionary theories, the laws of physics and the anatomical perception of the body-sex). Far from reiterating the critical impasse between the natural and the cultural—the realm of the given and the constructed—Deleuze and Guattari's conception of abstract matter or machine suggests an isomorphic method of analysis that maps the different yet connected levels of order of a body (the biophysical, the biocultural and the biodigital organizations of sex).

In this [essay], the process of endosymbiosis constitutes the abstract machine of sex or abstract sex. Abstract sex maps the isomorphic process of organization of different modes of information reproduction and communication. Lynn Margulis, the molecular biologist and theorist of endosymbiosis, or SET (serial endosymbiosis theory), explains how heterogeneous assemblages of molecules and compounds, unicellular and multicellular bodies, proliferating through gene trading, cellular invasion and parasitism, produce new cellular and multicellular compositions of bodies (Margulis 1981; Sapp 1994). In particular, merged bacteria that infect one another and symbiotic cellular associations reinvigorated by the incorporation of their contaminating diseases, map the potential mutation of bodies and sexes.

The Darwinian logic of evolution, resting on the centrality of sexual reproduction in order to engender species variations or differences, is substituted with a rhizomatic recombination of information expanding through viral hijacking of codes between singular machines of reproduction: a microbe and an insect, a bud and a flower, a toxin and a human. A far cry from organic unity and identity or from the original line of descent, endosymbiosis or abstract sex starts from heterogeneous assemblages where the parasiting web between hosts and guests produces new bodies-sexes. Far from determining a dualism between micro and macro levels of composition, for example between bacterial and nucleic cells, endosymbiosis exhibits a reversible feedback of information transfer that unfolds a continual variation of the body-sex, nature and matter.

This abstract machine provides a consistent method to analyze the manifold compositions of biophysical, biocultural and biodigital levels of modification of sex and reproduction. This isomorphic organization explains the dynamics of distinct machinic assemblages, cutting across micro and macro orders, and defines an immanent connection between bacterial sex and biodigital cloning, nucleic sex and disciplinary reproduction through singular points of mixture and differentiation of transmission. Abstract sex deploys the consistent relations between different machines of sex: from the autocatalytic association of cells to the association of multicellular bodies, from the society of bacteria to the social domain of disciplinary sex, from the digital culture of cloning images to the biotechnological proliferation of engineering cells. This consistency demarcates the autonomy of abstract sex—the endosymbiotic mutations of sex or desire—from the biological structures of the organic body and the cultural structures of signification, from the primacy of organic and linguistic totalities. It is not a matter of socio-cultural imitations of the natural or biological imitations of society. What comes first is neither a given essence nor the signification of essence. Rather, the abstract concatenation of bodies-sexes delineates the primacy of heterogeneous mixtures or symbiosis—biophysical elements, socio-cultural energies, economic trades, technical inventions, political forces and particles of desire—unfolding the potential of a body to become (mutate).

Instead of re-articulating sex within a post-feminist critical framework where difference is no longer material, abstract sex extends the feminist politics of desire by mapping the transversal mixing of information between bodies of all sorts (bacteria, vegetables, animals, humans, and technical machines). Abstract sex proposes to tap into the kinetic ethology of tiny sexes that lay out a micropolitics of symbiotic relations between different levels of mutation of matter and desire. The biophysical (the cellular organization of bacteria, eukaryotic cells and multicellular bodies), biocultural (the techno-scientific organization of the human body) and biodigital (the informatic manipulation of the human body) mutations of the body explain the entanglement of sex with sexual reproduction, the emergence of the two sexes, and the sex-gender association beyond the biological essence and the discursive construction of the body-sex.

It could be argued that this micropolitics exclusively highlights molecular differences or mutations of the body-sex by discarding, for instance, the feminist commitment and engagement with the macropolitics of representation that still determines the identity politics of sexual difference. Similarly, it might be observed that the microcosm of differences is not sufficient to account for body politics where categories of difference (gender, race and class) are still crucial for the situated conditions of minorities in global capitalism. Without dismissing these objections, abstract sex suggests that the micro levels of variation of the body (nature-matter) are crucial to produce a nonreductive understanding of difference (i.e. starting from zero or the plane of pure difference) in relation to the bio-technological engineering of cultures, bodies and life. The bio-technological mutations of the body point to the emergence of a micro level of difference proliferating through the symbiotic engineering of information crossing not only species and sexes, but also humans and machines. Far from abandoning difference, abstract sex connects biotechnological mutations to the mutations of desire announcing a new phase in the symbiotic becoming of the body-sex.

Abstract sex is a machinic concept that is not full of meanings, but is above all full of potential variations of the body. These variations emerge from a concatenation of small causes unleashing vast indirect effects that lead to a new conception and perception of sex. Concepts are operators of forces whose deployment is not related to the realm of possibilities, but to the plane of invention of a new kind of reality. Concepts have a political resonance, but this is not an immediate or direct one. Rather, they have to be continuously re-engineered in order to map the emergence of novelty. For this reason, the biotechnological disentanglement of sex from sexual reproduction is not to be reduced to the traditional dichotomy between biological conditions (embodiment) and techno-scientific discourses (disembodiment), but needs to be related to the connecting layers of organization (or stratification) of matter affecting bodies-sexes, societies, cultures and economies. [. . .] In order to engage with the new implications of the biotechnological mutations of the body, abstract sex argues that sex, far from being signified or represented, is primarily stratified.

Notes

1 D. J. Haraway argues that the convergence of biological and technological systems involves the emergence of a common language of codification produced by cybernetics. See Haraway 1991, 149–201; on the cyborg see also Gray 1995, Bell and Kennedy 2000, Kirkup et al. 2000, Flanagan and Booth 2002. [. . .]

2 Cyberfeminism has been influenced by Judith Butler's concept of "gender performance" (1993, 1990). Although I do not discuss Butler's concept of performativity, I will indirectly refer to this discursive understanding of sex.

3 On the critique of the economy of representation and the metaphysical notion of essence, see Deleuze 1994, 1990. On the critique of Plato's and Aristotle's essence in Deleuze, see De Landa 2002. See also Irigaray 1985.

4 The word "virtual" is derived from the Medieval Latin *virtualis*, itself derived from *virtus*, meaning strength or power. In scholastic philosophy the virtual is that which has potential rather than actual existence (Bergson 1991, 127–31, 210–11; Deleuze 1988a, 42–43, 55–62, 100–101).

Works Cited

Balsamo, Anne Marie. 1996. *Technologies of the Gendered Body: Reading Cyborg Women*. Durham: Duke University Press.

Bell, David, and Barbara M. Kennedy, eds. 2000. *The Cybercultures Reader*. New York: Routledge.

Bergson, Henri. 1991. *Matter and Memory*. Translated by N.M. Paul and W.S. Palmer. New York: Zone Books.

Braidotti, Rosi. 1994. *Nomadic Subjects: Embodiment and Sexual Difference in Contemporary Feminist Theory, Gender and Culture*. New York: Columbia University Press.

Butler, Judith. 1990. *Gender Trouble: Feminism and the Subversion of Identity*: Psychology Press.

———. 1993. *Bodies that Matter: On the Discursive Limits of "Sex"*. New York: Routledge.

De Landa, Manuel. 2002. *Intensive Science and Virtual Philosophy*. New York: Continuum.

Deleuze, Gilles. 1972. *Proust and Signs*. New York: G. Braziller.

———. 1988a. *Bergsonism*. New York: Zone Books.

———. 1988b. *Spinoza, Practical Philosophy*. San Francisco: City Lights Books.

———. 1990. *The Logic of Sense, European Perspectives*. New York: Columbia University Press.

———. 1994. *Difference and Repetition*. London: Athlone Press.

Deleuze, Gilles, and Félix Guattari. 1988. *A Thousand Plateaus: Capitalism and Schizophrenia*. Translated by Brian Massumi. London: Athlone Press.

Flanagan, Mary, and Austin Booth. 2002. *Reload: Rethinking Women + Cyberculture*. Cambridge, MA: MIT Press.

Foucault, Michel. 1977. "Theatrum Philosophicum." In *Language, Counter-memory, Practice: Selected Essays and Interviews*, 165–99. Oxford: Basil Blackwell.

Gatens, Moira. 1996. "Through a Spinozist Lens: Ethology, Difference, Power." In *Deleuze: A Critical Reader*, edited by Paul Patton, 162–87. Oxford and Malden: Blackwell.

Gray, Chris Hables, ed. 1995. *The Cyborg Handbook*. New York: Routledge.

Guattari, Félix. 1984. *Molecular Revolution: Psychiatry and Politics*. New York: Penguin.

Haraway, Donna. 1985. "A Manifesto for Cyborgs: Science, Technology, and Socialist Feminism in the 1980s." *Socialist Review* 80 (March–April 1985): 65–107.

———. 1991. *Simians, Cyborgs, and Women: The Reinvention of Nature*. New York: Routledge.

Hayles, N. Katherine. 1999. *How We Became Posthuman: Virtual Bodies in Cybernetics, Literature, and Informatics*. Chicago: University of Chicago Press.

Irigaray, Luce. 1985. *Speculum of the Other Woman*. Ithaca, NY: Cornell University Press.

Kirkup, Gill, Linda Janes, Kath Woodward, and Fiona Hovenden. 2000. *The Gendered Cyborg: A Reader*. New York: Routledge.

Margulis, Lynn. 1981. *Symbiosis in Cell Evolution: Life and its Environment on the Early Earth*. San Francisco: W. H. Freeman.

Plant, Sadie. 1997. *Zeros + Ones: Digital Women + the New Technoculture*. 1st ed. New York: Doubleday.

———. 1998. "Coming Across the Future." In *Virtual Futures*, 30–36. New York: Routledge.

———. 2000. "On the Matrix: Cyberfeminist Simulations." In *The Gendered Cybor: A Reader*, edited by Gill Kirkup, Linda Janes, Kath Woodward and Fiona Hovenden, 265–75. New York: Routledge.

Sapp, Jan. 1994. *Evolution by Association: A History of Symbiosis*. New York: Oxford University Press.

Springer, Claudia. 1996. *Electronic Eros: Bodies and Desire in the Postindustrial Age*. 1st ed. Austin: University of Texas Press.

Squires, Judith. 1996. "Fabulous Feminist Futures and the Lure of Cyberculture." In *Fractal Dreams: New Media in Social Context*, edited by Jon Dovey, 194–216. London: Lawrence & Wishart.

Stone, Allucquere Rosanne. 1991. "Will the Real Body Please Stand Up?" In *Cyberspace: First Steps*, edited by M. Benedikt, 81–118. Cambridge, MA: MIT Press.

Turkle, Sherry. 1995. *Life on the Screen: Identity in the Age of the Internet*. New York: Simon & Schuster.

Toxic Animacies, Inanimate Affections

MEL Y. CHEN

Toxic Allure

A toxin threatens, but it also beckons. It is not necessarily alive, yet it enlivens morbidity and fear of death. A toxin requires an object against which its threat operates; this threatened object is an animate object—hence potentially also a kind of subject—whose "natural defenses" will be put to the test, in detection, in "fighting off," and finally in submission and absorption.

This essay suggests that thinking, and feeling, with toxicity invites a recounting of the affectivity and relationality—indeed the bonds—of queerness as it is presently theorized. [. . .] I first consider how vulnerability, safety, immunity, threat, and toxicity itself are sexually and racially instantiated in the recent panic about lead content in Chinese-manufactured toys exported to the United States. [. . .] The essay ends by suggesting that the queering and racializing of material other than human amounts to a kind of *animacy*. Animacy is built on the recognition that abstract concepts, inanimate objects, and things in between can be queered and racialized without human bodies present, quite beyond questions of personification. Theorizing this animacy offers an alternative, or a complement, to existing biopolitical and recent queer-theoretical debates about life and death, while the idea of toxicity proposes an extant queer bond, one more prevalent today than is perhaps given credit. Such a toxic

Mel Y. Chen, "Toxic Animacies, Inanimate Affections," in *GLQ: A Journal of Lesbian and Gay Studies*, Volume 17, no. 2–3, pp. 265–286.

queer bond might complicate utopian imagining, as well as address how and where subject-object dispositions might be attributed to the relational queer figure.

Toxins—toxic figures—populate increasing ranges of environmental, social, and political discourses. [. . .] One recently crystallized metaphor points to a central culprit of the current global recession, and speaks precisely to this notional expansion of toxicity and its likely foray into its former history as a concept directly tied to immunity: "toxic assets." In this notion, asset is a good precisely because it entails capital value, but one which has unfortunately become—considering the discourse in which toxic asset has meaning—not only toxic but also perhaps "untouchable" (as an affective stance), "unengageable" (as tokens of exchange with limited commensurability), and perhaps even "disabling" (i.e., rendering the corporations that buy up those assets invalid themselves). The toxic assets of significance in the U.S. context, which are held responsible for global economic fallout, are the financial products composed of grouped mortgages tied to a hypervalued and/or unstable residential real estate market. Yet looking beyond financial products to other cultural sites, objects, or identities under capitalism, I suggest there are more toxic assets with which one might think economically, rhetorically, and in terms of critical domesticities. Given its rapidly multiplying meanings, toxicity clearly has a persistent allure. In what follows, I investigate the potential to resignify toxicity as a theoretical figure, in the interest of inviting contradictory play and crediting queer bonds already here: the living dead, the dead living, antisocial love, and inanimate affection.

Lead as Toxic Asset

Wrapped up in industrial manufacture and threatening "healthy development" with disability, the chemical element of lead has arguably become a "toxic asset." In the summer of 2007, lead became a primary concern precisely as a toxin in the U.S. media landscape. In this geopolitical and cultural moment, the critical scene was one of toys: lead's identity as a neurotoxic "heavy metal" was attributed to toys identified as made in China, toys whose decomposable surfaces when touched yielded up lead for transit into the bloodstreams of young children, giving it a means for its circulatory march toward the vulnerable, developing brain. Media outlets paraded images of plastic and painted children's toys as possibly lead-tainted and hence possible hosts of invisible threat. Medical professionals repeated, almost ritualistically, caveats about "brain damage," "lowered IQ's" and "developmental delay," directing their comments to concerned parents of vulnerable children. Toy testing centers were set up across the country.[1]

Journalists, government offices, and parents began to draw tighter connections between Chinese-made products and environmental toxins at large, and their lists now included heparin in Chinese-made medicines, industrial melamine in pet food, even Chinese smog, which had become unleashed from its geographic borders and was migrating to other territories. A generalized narrative about the inherent health risk of Chinese products (to U.S. denizens) began to crystallize.[2] Mass media pitched these environmental threats neither as "acts of God" nor as products of a global industrialization, but as invasive dangers into the U.S. territory from other national territories. These environmental toxins were supposed to be "there," but were found "here." [. . .]

The last few decades, particularly after 9/11, have seen a strengthened union of affects around terrorism that associate it with transnational provenance and hence invasive threat.[3] Under these conditions, the invisible threat of cognitive and social degradation in the case of lead meant that the abiding, relatively much more methodical, and diversified work of environmental justice activists on lead toxicity was here transmogrified into something that looked less "environmental" and more like another figure in the war on terror, a war that marked the diffuseness, unpredictability, and sleeper-cell provenance of enemy material and its biological vectors (Sze 2007; Calpotura and Sen 1994; Bullard 2005). This "war on terror" was doubly pitched as a neomissionary insistence on disseminating the "American way," including its habits of free choice and its access to a free market at its core defined by the proliferation of consumer products. Thus the very title of a *New York Times* article by Leslie Wayne about corrosive drywall for new homebuilding sourced from China—"The Enemy at Home"—betrays toxic drywall's coding as a biological threat metaphorized as war (itself not at great notional distance from "biological warfare") and/or as a symptomatic signifier of a war of capital flows (Wayne 2009). Lead, then, simultaneously became an instrument of heightened domestic panic, drawing from and recycling languages of "terror," and a rhetorical weapon in the rehearsal of U.S. economic sovereignty. [. . .]

The florid palette of toy-panic media representations yielded two prominent and repeated icons: the vulnerable child, more frequently a young, white, middle-class boy, and a dangerous painted toy, Thomas the Tank Engine. [. . .] Thomas the Tank Engine is a fetishized object, but not only of and for children: the series is marketed to middle-class parents who insist on high-status "quality" products, which in this case are aimed at boys and quite explicitly direct their proper masculine development. [. . .]

Displaced Racializations

Just as the presumed agents of "terror" have become racialized as Arab and/or Muslim after 9/11, so too has lead itself become recently racialized as Chinese. This particular racialization is a contemporary one. Before this transnationalizing of environmental threats, lead was, for example, a domestic concern in the United States, framed in terms of the public-health injunction to reduce the amount of leaded paint existing in older homes. In the late twentieth century, cultural media outlets like National Public Radio informed the liberal public that rates of lead poisoning among black children had much to do epidemiologically with the pollution of neighborhoods populated largely by people of color, given the existence of older buildings whose lead paint had not been remediated and the proximity of lead-polluting industrial centers.

Lately, however, the media identification of black children's vulnerability to the dangers of domestic lead has shifted dramatically in favor of identifying white children's vulnerability to the dangers of Chinese lead. In this potent narrative, black children have largely been superseded. I suggest it is not necessarily correct to judge that African American youth are now no longer viewed as vulnerable to lead. Rather, it is easier to imagine that in this pointedly transnational battle of sovereignty among major economic powers, black children are now the less urgent population under threat. It is, instead, as if black children are constructed as more proximate to lead itself, as naturalized *to* lead, new ground to the newest figure. A racial construction of blacks as already unruly, violent, contaminated, and mentally deficient lies inherent in the current neoliberal economy, itself an economic mode conditioned and supported by a growing and incredibly powerful prison-industrial complex with its own structuration of race, class, and gender (Davis 2003; Gilmore 2007). Lead exposure itself is associated with cognitive delay, enhanced aggression, impulsivity, convulsions, and mental lethargy. One wonders to what degree any newfound alarmism about the vulnerability of black children to environmental lead can succeed, given the abiding construction of the mentally deficient, impulsive, and spastic black body. That is to say, which assets have gone toxic (lead), which assets are considered toxic (bodies of color), which assets must be prevented from becoming toxic?

In the present case of the Thomas trains, then, lead toxicity is racialized, not only because the threatened future has the color of a white boy but also because that boy must not change color. The boy can change color in two ways: first, lead lurks as a dirty toxin, as a pollutant, and it is persistently racialized as anything but white. Second, the great fear of lead toxicity's neurological effect, borne out by toxicological evidence, is that lead makes a dull and/or violent child—it increases aggression and arrests some cognitive development. Some years ago, as I indicated above, before this domestic narrative largely disappeared in favor of the Chinese one, the greater public had been invited to simply extend a naturalized myth of decrepitude in urban blackness and hence imagine black children licking the peeling walls of their unmaintained dwellings as a decisive factor in black children's greater lead toxicity. This version of liberal environmentalism supports the progressive extension of "environmental rights" to previously unrecognized populations, yet does not critique environmental racism's structural makeup (Murphy 2006).[4] That is, black children are assumed to be toxic, and lead's threat to white children is not only that they risk becoming dull, or cognitively defective, but also that they lose their class-elaborated white racial cerebrality and become suited to living in the ghettoes.[5]

Lead Licking

The iconic white boy is an asset that must not be allowed to become toxic: he must not be mentally deficient, delayed, or lethargic. His intellectual capabilities must be assured to consolidate a futurity of heteronormative (white) masculinity, which is also to say that he must not be queer. I suggest here that one aspect of the threat of lead toxicity is its origin in a forbidden sexuality, for the frightening originary scene of intoxication is one of a *queer licking*. Here again is the iconic example of the white boy, who in the threatening or frightening scene is licking the painted train, a train whose name is Thomas, the train that is also one of the West's preeminent Freudian phallic icons.[6] This image never appears literally, or at least I have not seen it. Rather, if a boy and a train are present, the boy and the train are depicted proximately, and that is enough to represent the threat (the licking boy would be too much, would too directly represent the forbidden).

Precisely what is wrong with the boy licking the train? Many things are wrong: one, the boy licking Thomas the Tank Engine is playing improperly with the phallic toy, not thrusting it forward along the floor but putting it in his mouth. Such late-exhibited orality bears the sheen of that "retarded" stage of development known as homosexuality.[7] Thus "retarded," the scene slides further into queerness, as queer and disabled bodies alike trouble the capitalist marriage of domesticity, heterosexuality, and ability: the queer disability theorist Robert McRuer writes that the "ideological reconsolidation of the home as a site of intimacy and heterosexuality was also the reconsolidation of the home as a site for the development of able-bodied identities, practices, and relations" (2006, 88–89). Exhibiting telltale signs of homosexuality and lead toxicity alike is simultaneously to alert a protected, domestic sphere to the threat of disability. Finally, the queerest bioterrorist is one who is remote, racialized "otherwise," and hybrid: both human painting agents and microcosmic pollutants that, almost of their own accord, invade the body through plenitudes of microcosmic sites (a child's skin), sites the state cannot afford to acknowledge, for the queer vulnerabilities they portend.

The mediation of lead toxins in and around categories of life in turn undoes lead's deadness by reanimating it. In other words, like any toxin, lead has the capacity to poison definitively animate beings, and as such achieves its own animacy as the agent who can do us harm. To call it "personified" would be too simple. Toxins sometimes bear the threat of death to a protected life, but whether or not they "are" alive is not the issue. What is felt along with toxicity; what are its coextant biopolitical figures? [. . .] Common notions of toxicity invoke threatened immunity as their requisite condition. Immunity bears its own complex political histories;

Donna Haraway writes that immune systems are tightly intertwined with the biopolitical brokerages between "us" and "them" (Foucault 1978, 1970; Haraway 1989, 3–43; 1992, 295–337). An immune system is never innocent, never "merely" biological, because what is biological is itself never innocent of complex "intertextuality": scientific, public, and political cultures together inform understandings of the immune system. Haraway's politicization of the immune system is not surprising, because the medicalized notion of immunity was derived from political brokerages (Cohen 2009). Such knowledges comprising the "immune system" would seem, therefore, to serve as discourses that implicitly inform what is understood of the participants and as means of a perceived attack.[8]

What becomes of life when human bodies, those preeminent containers of life, are themselves pervaded by xenobiotic substances—that is, substances not intrinsic to, not generated by, unadulterated bodies (pollutants, synthetic pharmaceuticals, toxic heavy metals)—and nanotechnology? I suggest toxicity becomes significant now for reasons beyond the pressing environmental hazards that encroach on zones of privilege, beyond late-transnational capitalism doing violence to national integrities: debates about abortion and the lifeliness or deathliness of Terri Schiavo suggest not only that we cannot tell what is alive or dead, but perhaps that the diagnostic promise of the categories of life and death is itself in crisis.[9] [. . .][10]

Toxic Theory

Matters of life and death have arguably underlain queer theory from at least the time of its nomination in the early nineties, when ACT UP and radical queer AIDS activism blended saliently with the academic theorizing of politics of gender and sexuality. Signal to queer theory's interest in queer relationality, Lee Edelman (2004) takes up a psychoanalytic analysis of queerness's figural deathly assignment in relation to a relentless reproductive futurity. Jasbir Puar points to life-death economies that simultaneously segregate some queer subjects to the privileged realms of biopolitically "optimized life" while other perverse subjects are consigned to the realm of death, as a "result of the successes of queer incorporation into the domains of consumer markets and social recognition in the post-civil rights, late twentieth-century" (2007, xii). Similar affective pulses of surging lifeliness or morbid resignation might reflect the legacy of the deathly impact of AIDS in queer scholarship and might as well have reflexes in utopian or anti-utopian thinking in queer theory. Suggesting a "horizonal" imagining whose terms are pointedly not foretold by a pragmatic limitation on the present, José Esteban Muñoz (2009) offers a way around the false promise of a neoliberal utopia whose major concerns are limited to gay marriage and gay service in the military: lifely for a few, deathly for others.

Toxicity straddles boundaries of "life" and "nonlife," as well as the literal bounds of bodies, in ways that introduce a certain complexity to the presumption of integrity of either lifely or deathly subjects. While never undergoing sustained theorization in queer theory, toxicity has nevertheless retained a certain resonance there and a certain citational pull.[11] Roberto Esposito's *Bios* develops the idea of the "immunizing paradigm," which in his view is implicitly interwoven with community. Immunity is thus contracted on a "poisoned" affect of gratitude (on the basis of membership in a community) that undercuts the final possibility of individual immunity. Imbalances are inherent to the model; an "interdependent social ecology of bodies" could easily yield desires for greater protection, and some bodies might legally build greater immunity against others (2008, xiii). Esposito identifies the shaky prescription of the introjection of the negative agent as a way to defend against its exterior identity. I wonder, however, whether toxicity meddles with the subject-object relations required for even this immunitary ordering that Esposito suggests. Who is, after all, the subject here? What if the object, which is itself a subject, has been substantively and subjectively altered by the toxin? At the same time, toxicity releases "life" from any absolute need to contain or protect it. Toxicity is simultaneously released from the realm of the dead, even as immunity remains premised on the generativity of life.[12]

I find myself dancing in this essay between advocating the notional release of the metaphor of toxicity and marking its biopolitical entrainment as an instrument of difference. While the first seems theoretically important to allow a kind of associative theorizing, it is important to retain simultaneously a fine sensitivity to the vastly different intersectional sites in which toxicity involves itself in very different lived experiences (or deaths)—for instance, a broker's relation to "toxic bonds" versus a farmworker's relation to pesticides. One toxin is metaphorical, the other literal. Yet metaphorical luxuries can have deadly consequences. Michael Davidson reminds us that while literary analogical treatments of disability render disabled characters as functional prostheses who are merely there to help entrench a nondisabled subject position, "there are cases in which a prosthesis is *still* a prosthesis" (1997, 120). Sometimes a mask is still a mask, even if it is simultaneously a masquerade.

Animacy, Interobjectivity

A discussion of toxicity and affect calls for a concomitant discussion of the idea of *animacy*. Sianne Ngai (2005) demonstrates how one of animacy's correlates, animatedness, can become a quality of racialized affect. Yet the word *animacy* has no single definition. It is described alternately as a quality of agency, sentience, or liveness; it is also a term of linguistic semantics that registers the grammatical ramifications of the sentience of a noun. It can also be considered a philosophical concept that addresses questions of life and death. These many meanings must be sustained together, for they all circulate biopolitically, running through conditionally sentient and nonsentient, live and dead, agentive and passive bodies. We can then ask not "who is alive, or dead," but "what is animate, or inanimate, or less animate"; relationally, we can ask about the possibilities of the interobjective, above and beyond the intersubjective.[13] For instance, Jennifer Terry's recent work on the love of objects, as well as the tradition of fetish scholarship, speaks to an intensified investment in objects; it is useful to build on this work, then, to ask questions of the subject facing that object, precisely how or why to mark its subjectivity as such, and when instead to consider its objectivity (2009). This interobjective tack is suggested, for instance, by the above example of the couch, with which my relationality is made possible only to the degree that I am not in possession of human sociality.

Sara Ahmed writes extensively about her orientation toward a table of hers and that table's orientation toward her. "We perceive the object as an object, as something that 'has' integrity, and is 'in' space, only by haunting that very space; that is, by co-inhabiting space such that the boundary between the co-inhabitants of space does not hold. The skin connects as well as contains. . . . Orientations are tactile and they involve more than one skin surface: we, in approaching this or that table, are also approached by the table, which touches us when we touch it" (2006, 54). I first must agree, but then find that what she nevertheless still presumes in this work is the proper integrity of her body and of the table, an exclusion of molecular travel that permits her to position one thing against another. Yes, she is talking mainly about the perception of integrity, but my contention here is that percepts are to some degree bypassed, for instance, by the air itself. Standing before you, I ingest you. There is nothing fanciful about this. I am ingesting your exhaled air, your sloughed skin, and the skin of the tables, chairs, and carpet in this room.

Ahmed's reading takes for granted the deadness and/or inanimacy of that table, as a reference point for the orientation of a life, one in which the table is moved according to its owner's purposes and conveniences. And while it would be unfair to ask of her analysis something not proper to its devices, I do wonder how this analysis must change once the animate/inanimate object distinctions collapse, when we move beyond the exclusionary zone made up of the perceptual operands of phenomenology. The affective relations I have with this couch are not made out of a predicted script and are received as no different from those with animate beings, which, depending on perspective, is both their failing and their merit. My question here is, what is lost when we hold tightly to that exceptionalism that says that couches are dead and we are alive? For would not my nonproductivity, my nonhuman sociality, render me some *other* human's "dead"—as certainly it has, in case after case of the denial of disabled existence, emotional life, sexuality, or subjectivity? Or must couches be cathected differently from humans? Or do only certain couches deserve the attribution of a (sexual) fetish? These are only questions to which I have no ready answers, except to declare that those forms of exceptionalism no longer seem reasonable.

For animacy is a category mediated not by whether you are a couch, a piece of lead, a human child, or an animal but by how you interpret the thing of concern and how dynamic you wish it to be. Above and beyond the philosophical intersubjectivity we might analytically afford ourselves, there is the strict physicality of the elements that travel in, on, and through us, and sometimes stay. If we ingest each other's genetic code-driven replication of skin cells, as well as each other's personal care-driven application of synthetic skin creams, then animacy comes to appear as a category itself held in false containment. Also, the toxicity of the queer to the heterosexual collective or individual body; the toxicity of the dirty subjects to the white empire; the toxicity of heavy metals to an individual body: none of these segregations perfectly succeeds even while it is believed with all effort and investment to be effective.

In perhaps its best versions, toxicity propels, not repels, queer loves, especially once we release it from exclusively human hosts, disproportionately inviting disability, industrial labor, biological targets—inviting loss and its "losers," and trespassing containers of animacy. We need not assign the train-licking boy so *surely* to the nihilistic underside of futurity or to his own termination, figurative or otherwise. I would of course be naive to imagine that toxicity stands in for utopia, given the explosion of resentful, despairing, painful, screamingly negative affects that surround toxicity. Nevertheless, I do not want to deny the queer productivity of toxins and toxicity, quite beyond the given enumerable set of addictive or pleasure-inducing substances, or to neglect indeed to ask after the desires, the loves, the rehabilitations, the affections, the assets that toxic conditions induce. Unlike viruses, toxins are not so very containable or quarantinable; they are better thought of as conditions with effects, bringing their own affects and animacies to bear on lives and nonlives. If we move beyond the painful "antisocial" effects to consider the sociality that is present there, we find in that sociality a reflection on extant socialities among us, the queer-inanimate social lives that exist beyond the fetish, beyond the animate, beyond the pure clash of human body sex.

Notes

1 For a more detailed account of the lead panic and its shared resonances with the peculiar toxicity of a much earlier Fu Manchu fantasy about Asian threat materialized in the form of an interspecies/inanimate "serum," see Chen 2007, 367–83.

2 The actual picture is dramatically more complex. Chinese residents are being poisoned by their "own" industries, through pollution of water, air, food, and soil, and the regular failure of government protections from industrial toxins has led to a dramatic rise in community protests, lawsuits, and organized activist movements.

3 Nonstatehood has come into mature relationship with the possibility of terrorism, evidenced most recently by the fact that U.S. Senator Joe Lieberman has, with some support, proposed revoking the citizenship of those

who demonstrate financial support or other forms of allegiance to U.S.-deemed "terrorist" organizations.

4 For the class and race qualifications of the sick building syndrome movements and forms of activism, see Murphy, 2006.

5 "A mind is a terrible thing to waste," reads the United Negro College Fund's campaign to further blacks' access to education. Dan Quayle's perversion of this slogan, "What a terrible thing it is to lose one's mind," suggests what fantasies about blackness might underlie benevolent white representations.

6 I thank Don Romesburg for first getting me to indulge in this sensory fantasy.

7 I am invoking the impossible juncture between the queernesses "naturally" afforded to children and the fear of a truly queer child (Stockton 2009; Bruhm and Hurley 2004).

8 For more extensive studies of immunity (which toxicity implicates), see Esposito 2008; Cohen 2009; Martin 1994.

9 Giorgio Agamben's *Homo Sacer: Sovereign Power and Bare Life* (Agamben 1998) has had a tremendous impact on the thinking of the politicized line and relationship between life and death. Achille Mbembe extends this theorizing into postcolonial modes of analysis (2003, 2001).

10 *Editor's note:* See the original article for the sections "Toxic Sensorium," "Queer Ingestion," and "Intoxicated Subjects"

11 Eve Sedgwick's implicit logic of the toxic as an excisable element of a self and her concomitant rejection of toxicity as a model for shame was in some ways redeployed with a difference by Muñoz, who used the notion of disidentification to represent the willing uptake of toxic elements in order to pose new figurations of identity and minoritarian/majoritarian politics. Christine Bacareza Balance juxtaposes public health's indictment of queer Filipino bodies with disproportionately high HIV rates as "toxic subjects" with the possibility of shared queer Filipino American drug trips as pleasurable and intimate counterpublics (Sedgwick and Frank 2003; Muñoz 1999; Balance 2006).

12 Thinking more specifically about the ethical/affective politics of geopolitical strife, particularly war, Judith Butler (2004) writes of vulnerability as a given condition, a condition that might inform a radically changed ethics were it to be acknowledged.

13 Thanks to Michael Israel for naming this investment.

Works Cited

Agamben, Giorgio. 1998. *Homo Sacer: Sovereign Power and Bare Life*: Stanford University Press.

Ahmed, Sara. 2006. *Queer Phenomenology: Orientations, Objects, Others*: Duke University Press.

Balance, Christine Bacareza. 2006. "On Drugs: The Production Of Queer Filipino America Through Intimate Acts of Belonging." *Women & Performance: A Journal of Feminist Theory* 16 (2): 269–281.
Bruhm, Steven, and Natasha Hurley. 2004. *Curiouser: On the Queerness of Children*. Minneapolis: University of Minnesota Press.
Bullard, Robert D. 2005. *The Quest for Environmental Justice: Human Rights and the Politics of Pollution*. 1st ed. San Francisco: Sierra Club Books.
Butler, Judith. 2004. *Precarious Life: The Powers of Mourning and Violence*. New York: Verso.
Calpotura, Francis, and Rinku Sen. 1994. "PUEBLO Fights Lead Poisoning." *Unequal Protection: Environmental Justice and Communities of Color*: 234–55.
Chen, Mel Y. 2007. "Racialized Toxins and Sovereign Fantasies." *Discourse* 29 (2): 367–383.
Cohen, Ed. 2009. *A Body Worth Defending: Immunity, Biopolitics, and the Apotheosis of the Modern Body*. Durham: Duke University Press.
Davidson, Michael 1997. "Universal Design: The Work of Disability in an Age of Globalization." In *The Disability Studies Reader*, edited by Lennard J. Davis, 117–130. New York: Routledge.
Davis, Angela Y. 2003. *Are Prisons Obsolete?* New York: Seven Stories Press.
Edelman, Lee. 2004. *No Future: Queer Theory and the Death Drive*. Durham: Duke University Press.
Esposito, Roberto. 2008. *Bios: Biopolitics and Philosophy*. Minneapolis: University of Minnesota Press.
Foucault, Michel. 1970. *The Order of Things: An Archaeology of the Human Sciences*. New York: Vintage Books.
———. 1978. *The History of Sexuality, Vol 1: An Introduction*. 1st American ed. New York: Pantheon Books.
Gilmore, Ruth Wilson. 2007. *Golden Gulag: Prisons, Surplus, Crisis, and Opposition in Globalizing California*. Berkeley: University of California Press.
Haraway, Donna. 1989. "The Biopolitics of Postmodern Bodies: Determinations of Self in Immune System Discourse." *Differences* 1 (1): 3–43.
———. 1992. "The Promise of Monsters: A Regenerative Politics for Inappropriate/d Others." In *Cultural Studies*, edited by Lawrence Grosberg, Cary Nelson and Nelson Treichler, 295–337. New York: Routledge.
Martin, Emily. 1994. *Flexible Bodies: Tracking Immunity in American Culture from the Days of Polio to the Age of AIDS*. Boston: Beacon Press.
Mbembe, Achille. 2001. *On the Postcolony*. Berkeley: University of California Press.
———. 2003. "Necropolitics." *Public Culture* 15 (1): 11–40.
McRuer, Robert. 2006. *Crip Theory: Cultural Signs of Queerness and Disability*. New York: NYU Press.
Muñoz, José Esteban. 1999. *Disidentifications: Queers of Color and the Performance of Politics*. Minneapolis: University of Minnesota Press.

———. 2009. *Cruising Utopia: The Then and There of Queer Futurity*. New York: NYU Press.

Murphy, Michelle. 2006. *Sick Building Syndrome and the Problem of Uncertainty: Environmental Politics, Technoscience, and Women Workers*. Durham: Duke University Press.

Ngai, Sianne. 2005. *Ugly Feelings*. Cambridge, MA: Harvard University Press.

Puar, Jasbir K. 2007. *Terrorist Assemblages:Homonationalism in Queer Times*. Durham: Duke University Press.

Sedgwick, Eve Kosofsky, and Adam Frank. 2003. *Touching Feeling: Affect, Pedagogy, Performativity*. Durham: Duke University Press.

Stockton, Kathryn Bond. 2009. *The Queer Child, or Growing Sideways in the Twentieth Century*. Durham: Duke University Press.

Sze, Julie. 2007. *Noxious New York: The Racial Politics of Urban Health and Environmental Justice*. Cambridge, MA: MIT Press.

Terry, Jennifer. 2009. "Objectum-Sexuality." Rethinking Sex: A State of the Field Conference in Gender and Sexuality Studies, University of Pennsylvania.

Wayne, Leslie. 2009. "The Enemy at Home." *New York Times*, October 8, 2009.

Plasmodial Improprieties

Octavia E. Butler, Slime Molds, and Imagining a Femi-Queer Commons

AIMEE BAHNG

> I consider myself a creature of the mud, not the sky.
>
> DONNA HARAWAY

WHEN FEMINIST-QUEER SCIENCE STUDIES LOOKS FOR ALTERNATIVE models for being in the world that move beyond the human, we would do well to consider the work of African American science fiction writer Octavia E. Butler,[1] who dedicated her life to imagining worlds otherwise through the generic medium of science fiction.[2] This paper posits Butler as a black feminist philosopher of science, who used the genre of speculative fiction to formulate nonhierarchical socialities and even more radical onto-epistemological modes of living in common, often through feminist ideas of collaborative praxis and queer notions of kinship.

Drawing on my archival research of Octavia Butler's collected papers at the Huntington Library, I point to Butler's unpublished research notes on slime molds and other nonhuman organisms as an example of thinking beyond the human prior to the more recent turn to new materialisms. Butler's approach to slime molds and what she learns from them, I argue, model modes of engagement with other life-forms that come from practiced thinking with alien-human entanglements. While Butler has emerged as one of the most celebrated black feminist science fiction writers in the world, in this paper I argue that the imaginative possibilities her writing and research practices engender constitute an example of feminist scientific inquiry we could call speculative fabulation. Fabulation spans the space between what speculative realists tend to position diametrically as the sheer ideation of the linguistic turn and the realism of matter (Bryant, Srnicek, and Harman 2011, 3; Meillassoux 2008, 5). It demands of its practitioners what Sara Ahmed might characterize as queer disorientation (2006). I interject Butler as a thinker who anticipates many of the recent critical moves beyond the human in feminist and queer theory, and I posit her literary works as theoretical interventions to these conversations that take into consideration histories of empire and slavery as phenomena at the planetary scale.

To begin, I focus on Butler's encounter with slime molds and how she begins to think about alternative ontologies and systems of organizing. Highlighting Butler's extrapolations from slime mold behavior to explore alien, human, and alien-human relations in her speculative fiction, I argue that Butler's fabulation of "xenogenesis" in her eponymous trilogy models an openness to the uncertain movements beyond the human that nonetheless foreground and stay attuned to power imbalances that too often narrow the possibilities of becoming. While Butler's thought experiment could be put into conversation with what Donna Haraway (2016) and Karen Barad (2007) respectively call "sympoiesis" or "intra-action," Butler's tale of xenogenesis suggests a deep imbrication of colonial modes of acquisition and genetic engineering as a science we have come to know in the US within the context of a capitalist, entrepreneurial mode of scientific research. Through a reading of Butler's fictional construct, I chase the implicit question: What would a feminist, decolonial genomics look like? Situating slime mold as a recurring player in feminist-queer science studies, I put Butler's research in conversation with Evelyn Fox Keller's work on slime mold reproduction and movement from the late 1960s through 1983. Together, Butler's notes and fiction provide a rich, alternative archive for feminist-queer science studies to examine as the field continues to focalize collaborative and collectivist frameworks for conducting science queerly.

But it won't all be utopian praise for slime mold. Starting from a moment of archival discovery, of thinking across time and space with Octavia Butler, this essay moves through some initial excitement about Butler's interspecies thinking to consider the more recent hype around and instrumentalization of slime mold in popular science as well as in speculative realist scholarship. While Butler's research into slime mold and other colony organisms indicates her interest in models of collective action, decentered modes of self-organizing, and systems of collaborative production, slime mold becomes, in the era of financialization and its attendant fields of probabilization and preemption, subject to more predatory forms of speculation. Swept up into a culture of optimization and risk aversion that celebrates its efficiency rather than its queerer characteristics, slime mold gets oriented toward models of competition when entrepreneurial technoscience asks it to perform spectacularized performances of problem-solving efficiency and adaptability.

In the final moves of the paper, I return to Butler and the slime mold, demonstrating how, despite her interest in its resistance to the atomizing proclivities of property, propriety, and privatization, she curbs her enthusiasm for colony organisms with a wariness around all-too-human systems of power that might confuse "emergence" for "colonization." Out of Butler's trepidation, I argue for the importance of keeping decolonial thought a part of feminist new materialist inquiry. Butler understands, on the one hand, that differentiation can fuel capitalist operations by cultivating the conditions for competition, and yet, on the other hand, that complete disregard for difference too often obfuscates power dynamics already in play even in the sympoietic moment. But first, let's join Octavia at the moment when she begins thinking about slime molds and other colony organisms.

The Impropriety of Social Amoebae

In Box 83, Folder 1625 of the Octavia E. Butler papers, housed at the Huntington Library in San Marino, California, a single note about slime molds surfaces (see opposite). Dated December 31, 1988, the note generally catalogs a number of colony organisms, such as the Portuguese man-o-war and the anglerfish. In multicolored pen on a lined index card, Butler has written: "We find true colony organisms rare and facinating [sic]. Here they are the exception[.] There, perhaps, the rule."

What is the "there" to which she refers? Not the soil through which slime molds travel (up to one centimeter per hour!), nor the sea depths where the female anglerfish "might carry more than one male" on her back. It is an elsewhere, a speculative space where someone—in this case, perhaps the most treasured black feminist speculative fiction writer of all time—can begin to imagine an otherwise. If "here" references a world

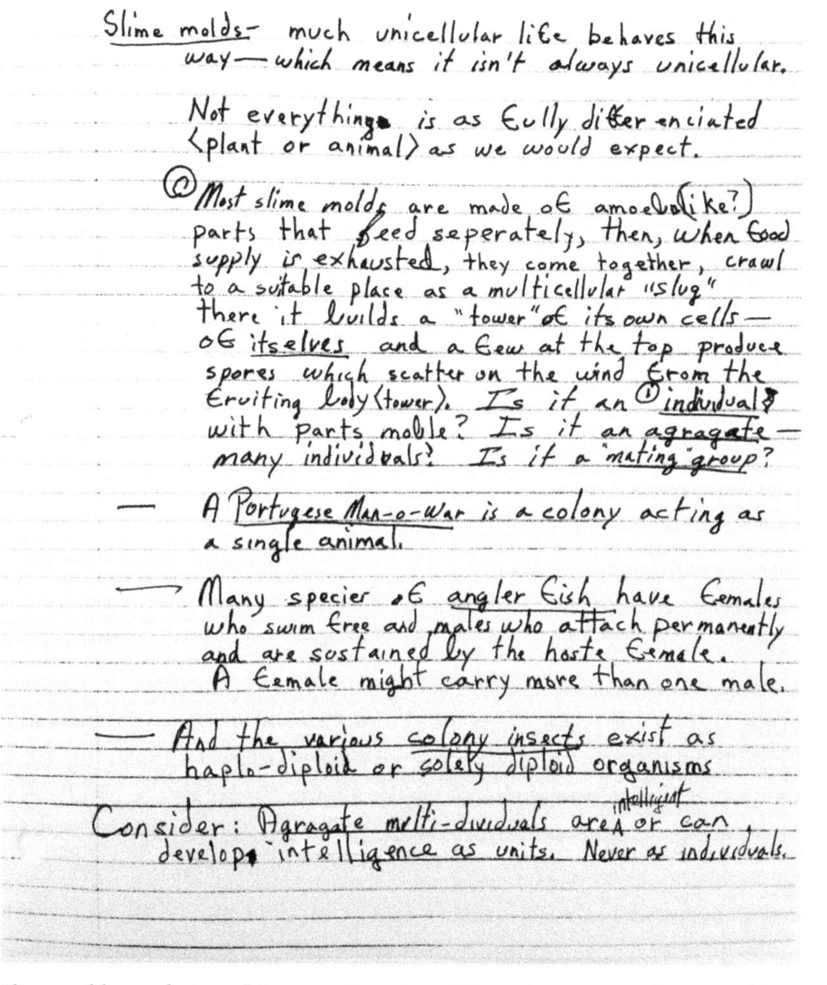

12-31-88

We find true colony organisms rare and facinating
Here they are the exception
There, perhaps, the rule.

Slime molds- much unicellular life behaves this way—which means it isn't always unicellular.

Not everything is as fully differenciated <plant or animal> as we would expect.

① Most slime molds are made of amoeba(like?) parts that feed seperately, then, when food supply is exhausted, they come together, crawl to a suitable place as a multicellular "slug" there it builds a "tower" of its own cells—of itselves and a few at the top produce spores which scatter on the wind from the fruiting body (tower). Is it an ① individual? with parts mobile? Is it an agragate—many individuals? Is it a "mating" group?

— A Portugese Man-o-War is a colony acting as a single animal.

→ Many species of angler fish have females who swim free and males who attach permanently and are sustained by the host female. A female might carry more than one male.

— And the various colony insects exist as haplo-diploid or solely diploid organisms

Consider: Agragate multi-dividuals are intelligent or can develop intelligence as units. Never as individuals.

Slime mold speculations. "Notes on Organisms," December 31, 1988. Box 83, Folder 1625, Octavia E. Butler Papers, Huntington Library, San Marino, California.

processed through the hegemonic filters of what some may call human civilization, Butler's "there" gestures toward other worlds: of slime molds and anglerfish, of organisms that belie taxonomic kingdoms, of life-forms and lifeways that elude our current frameworks. This note on slime molds, I contend, documents queer feminist science (fiction) in the making.

On slime molds specifically, Butler's note focuses on their queerness:

> Slime molds– much unicellular life behaves this way—which means it isn't always unicellular. . . .
>
> Most slime molds are made of amoeba(like?) parts that feed seperately [sic], then, when food supply is exhausted, they come together, crawl to a suitable place as a multicellular "slug[.]" [T]here it builds a "tower" of its own cells—of itselves[—]and a few at the top produce spores which scatter on the wind from the fruiting body <tower>. Is it an agragate [sic]—many individuals? Is it a "mating" group? (Butler 1988, emphasis in original)

Essentially an undifferentiated sack of multinucleated protoplasm, the cellular slime mold *Dictyostelium discoideum* has no brain, no central nervous system—and yet, in conditions of scarcity, it will swarm, intelligently reconfiguring itself into multicellular masses, working in tandem temporarily to proliferate, spread, and relocate to more generative sites. The slime mold defies Linnean taxonomization, as it cannot be easily categorized as animal, plant, mineral, or even fungi, leaving contemporary scientists to relegate the hundreds of species of slime molds to kingdom Protista, a kind of catchall kingdom of "others." Unsettling scientific classification, the slime mold even belies strict adherence to grammatical rules. In writing about slime mold, one can slip between singular and plural forms at every reference with due cause, as both cellular and plasmodial slime molds exist alternately as singular and plural, depending on how and when you're counting. Wondering whether slime mold is best characterized as an aggregate of individuals, a mating group, a swarm, or a single organism, Butler meets the question of pronouns with an admirable openness, queering and querying the limiting politics of either individualism or collective action. Describing the fruiting body as "a 'tower' of its own cells—of itselves," Butler bends grammar to accommodate this alien ontology, asserting the organism's nonconforming, decentralized organization. Butler's methods constitute queer science studies approaches. By fully recognizing the alien possibilities of this life-form—by insisting that not all unicellular life is always unicellular, and by meeting slime mold morphology in between singular and plural in its grammar—Butler demonstrates a remarkable openness to non-normative biological organization. She does not look to figure the slime mold out. She seems excited to follow it off the script of 1980s evolutionary biology to other possibilities. In slime, she looks for a model of life that could be, rather than life that already is. It is a speculative fabulation, drawn from life unruly.

Butler's inquiries into slime molds and what she calls "multi-dividual units" coincide with some of the key questions she raises around human-alien relations as well as nonhierarchical social structures in her three novels *Dawn* (1987), *Adulthood Rites* (1988), and *Imago* (1989), which comprise the so-called Xenogenesis trilogy, collected in 2000 in a single volume titled *Lilith's Brood*. Descriptions of slime mold behavior often focus on its anomalous self-organizing, which requires systemic morphing between single-celled and multicellular forms:

> *Dictyostylium* has the remarkable property of existing alternatively as single cells or as a multicellular organism. As long as there is enough food around, the single cells are self-sufficient, growing and dividing by binary fission. But, when starved, these cells undergo internal changes that lead to their aggregation into clumps which, as they grow bigger, topple over and crawl off as slugs. (Keller 1983, 516)

The transformation of "self-sufficient" cells into aggregated clumps and slugs could well describe the bodies of the Oankali, the alien species depicted in Octavia Butler's Xenogenesis series. The Oankali, who arrive at a postapocalyptic Earth and "save" a small group of humans for the potential of their genetic material, are covered in head and body tentacles that function as sensory organs. In times of stress, they knot up into clumps. One might also recognize slime mold chemotaxis in the walls and floors of the Oankali ship, which Butler describes as a living organism that digests and recycles its inhabitants' waste and communicates with them through biochemical signatures and feedback loops. Indeed, Butler has often fabulated species that embody symbiogenesis, which highlights cooperation rather than competition in describing the organization and evolution of complex life (Ferreira 2010; Vint 2010).

In Butler's fictional world, acclimating to this alien ontology requires an active queering of human sexuality vis-à-vis the third-gender "ooloi" of the Oankali. The ooloi anchor the mating ecologies among male, female, and non-Oankali participants who enjoy the benefits of genetic therapy and chemically stimulated pleasure. Lilith, who joins an Oankali family with an ooloi named Nikanj, helps Nikanj undergo the "internal changes" that humans might associate with puberty. Like a slime mold undergoing its transformation from unicellular to multicellular organism in a time of stress, Nikanj finds temporary relief in foraged food: "It drew its head and body tentacles into knots," Butler writes. "'Give me something else to eat.' [Lilith] gave it a papaya and all the nuts she had brought in. It ate them quickly. 'Better,' it said. 'Eating dulls the feeling sometimes'" (Butler [1987] 1997, 103). In fabulating the Oankali, Butler has drawn much from what could be considered slime mold's queerest properties: nondimorphic sexuality, trans-species chemo-tactile communication, and nonhierarchical sociality. In these ways, slime mold behavior itself speaks to femi-queer notions of collectivity and nonhierarchical social formations. Remarkably, researching slime mold behavior also leads directly to the very heart of feminist science studies in its emergence as a field.

In 1969, feminist physicist Evelyn Fox Keller, along with mathematician Lee Segel, looked to the slime mold as a demonstrable example of spontaneously emergent, self-organizing principles. Their preliminary research, though, was largely abandoned by the scientific brotherhood in favor of the so-called "pacemaker hypothesis," which suggested that a centralized authority, composed of special pacemaker or "founder cells," ordered other cells to aggregate. Despite the complete lack of evidence for the existence of such cells, the pacemaker hypothesis was upheld as conventional scientific knowledge throughout the sixties and seventies. In 1983, though, Keller definitively overturned this hypothesis with the help of developments in mathematical biology, including the study of nonlinear reaction-diffusion equations, which provided a means of understanding the interaction between the production and diffusion of acrasin and cellular chemotaxis. Chemotaxis, Keller revealed, *not* special founder cells, directs slime mold aggregation and movement. In her article, Keller exposes the extent to which scientists had imposed hierarchical and ultimately patriarchal structures of thinking onto cellular slime mold. To "posit a single central governor," she writes, was to subject scientific inquiry to a "zealous desire for familiar models of explanation, . . . imposing on nature the very stories we like to hear" (1983, 521).

Though many scientists sheepishly admit enjoying science fiction, many often disavow any significant influence cultural texts might have on the work they do in the laboratory, despite the common emphases on speculation and experimentation shared by scientists and science fiction writers alike (Haraway 1991; Milburn 2010; Shaviro 2016; Bahng 2017). Feminist science studies scholar Banu Subramaniam has called for "more engaging plots and stories that are located in the interdisciplinary fissures of the sciences and the humanities" (2014, 72).

At the conjuncture of science and fiction, Octavia Butler's speculative fabulation instantiates just such an assemblage of transdisciplinary knowledge making. Reading Butler's speculative fiction alongside scientific research on slime molds, one can begin to trace the entangled fictional and nonfictional stories of how human and nonhuman species organize themselves. One can begin to track the narrativization of human exceptionalism in the conventional story of life itself. And because slime molds lead us away from systems of hierarchical ordering, the story of how humans have tried to shoehorn slime into a more familiar form reveals how storytellers of science become susceptible to their own frameworks. In other words, while there may very well be a slime mold ontology beyond human understanding, one ethical way to reach across to that speculative reality might be to *wonder with* it, rather than *marvel at* it from a distance. In this way, considering Butler's work moves the new materialist conversation from trans-species allyship to multispecies solidarity, and in so doing, advances a feminist queer materialism as threaded through cross-ethnic antiracist work. Such consideration puts Butler's fabulations and Evelyn Fox Keller's research on slime mold aggregation in a more capacious feminist genealogy of nonhierarchical organizing that might include, for example, Jasbir Puar's theorization of political assemblage (2007), or Occupy, or #BlackLivesMatter theories of decentralized and nonhierarchical organizing.

Butler's study of the slime mold's transversal movement across and through single- and multicellular identities challenges notions of propriety, the proper, and the proper noun: She crafts the particularly queer pronoun "itselves" to describe slime mold differential collectivity. Slime molds organize themselves somewhat spontaneously and collectively. As Steven Shaviro describes it, the slime mold is "a *collective* without individuals, without any specialized parts, and without any sort of articulated (or hierarchical) structure" (2016, 195). Also called "social amoebae," slime molds, with their distributed modes of organization, constitute a radical departure from hierarchical organizational systems and also confound notions of privatization. Butler spent most of her time in public spaces—in public libraries and on public transportation. Indeed, her dyslexia made her nervous about driving, so the bus became a way for her to navigate the LA sprawl while also affording her the time-space in which to imagine the world in ways that transected the rather segregated neighborhoods and logics of privatization rapidly engulfing much of the Southland into racial and class enclaves.[3] Most of Butler's scientific research and thinking happened during her hours commuting on the bus to her various factory and temp jobs, or during her frequent trips to the Central Library. Even in 1988, at the accelerated turn of science into private funding, Butler was taking science back to public spaces.

Written on New Year's Eve, Butler's slime mold note falls at the cusp of multiple transitions. For one, 1988 is when she was wrapping up the Xenogenesis series and moving her thinking toward the Parable series and what would become a religious fabulation called Earthseed. The plasmodial improprieties that slime molds enact through channels of connectivity might also remind readers of Butler's grappling with notions of private gated communities and alternative possibilities for communal living, as well as Lauren Olamina's hyperempathy syndrome, from the Parable series. The timing of Octavia Butler's research on slime molds also coincides with the pinnacle of Reagan- and Thatcher-era financialization, deregulation, and privatization. The late '80s is precisely the era in which we see the financialization of science in particular, when, as Melinda Cooper has demonstrated, venture capitalists started funding scientific research largely based on its promise of deliverable goods that could be sold to a consumer culture being trained toward constantly upgradeable selves (2011). This form of speculation produces probable states as calculable outcomes in investment contracts (futures, options, swaps) and choices for individual portfolios (Bahng 2017). Such packaging *forecloses* alternative possibilities in the interests of a precise rate of return. Butler's speculations are more creative (Bahng 2017). They learn to learn from other human and nonhuman actors. They don't abide the proprietary norms of intellectual production in the era of the corporate university. No silos. No atomization. Just concatenation.

In slime mold–Oankali-Earthseed aggregation, I contend, Butler begins to experiment with forms of communing perhaps most akin to feminist Marxist formulations. Silvia Federici, for example, proposes a commons that exceeds human social sortings: "Indeed, if communing has any meaning, it must be the production of ourselves as a common subject [itselves]. This is how we must understand the slogan 'no commons without community.' But 'community' has to be intended not as a gated reality, a grouping of people joined by exclusive interests separating them from others, as with communities formed on the basis of religion or ethnicity, but rather as a quality of relations, a principle of cooperation and responsibility to each other and to the earth, the forests, the seas, the animals" (2012). But Federici's move away from communities of humans toward a set of relations among humans, animals, and the environment seems to propose a moving beyond "the subject" that fails to consider processes of subjection. In Butler's *Parable of the Sower* (1993), the gated community to which Federici gestures in this quotation clearly does have its limitations. In the near-future world in which Lauren Olamina founds Earthseed, the gated community is a failed remnant of private interests, but Earthseed, which replaces it, remains conflicted with very human forms of power. It is no utopia.

Decolonizing *Physarum polycephalum*

Slime molds have been made much of in recent popular science news headlines, as everyone from computer scientists to city planners began modeling the adaptive behavior of *Physarum polycephalum*—not a cellular but a plasmodial slime mold (aka myxomycete)—as part of a turn toward more complex, algorithmic methods for prediction and speculation. When presented with oat flakes arranged in the pattern of Japanese cities around Tokyo, *Physarum polycephalum* constructed networks of nutrient-channeling tubes that were strikingly similar to the layout of the Japanese rail system (Sanders 2010).[4] The telecom industry, which increasingly relies on so-called "emergent software" to plan how to lay down subterranean cable infrastructures most efficiently and with minimal disruption, has also turned to slime mold–based modeling, as the plasmodial organism lays down not only efficient pathways but also networks that stand the least chance of disruption should one strand be compromised or temporarily severed (Gorby 2009; Keim 2008). The plasmodial slime mold has become such a key modeling agent in commercial and scientific research that it has been used to "grow a computer" and was part of an experiment to predict Mexican migration patterns across the US (Adamatzky and Martinez 2013). As of 2014, slime molds are even now being bred and raced for entertainment (Hotz 2014).

While slime molds may offer some alternative to ways of organizing, there is reason to pause the celebration of the liberatory potential of the social amoebae. Innovators and entrepreneurs have folded slime molds into the workforce as experimental bodies, picked up for their efficiency and utility, but not for their queerness. If we hear an echo of the Oankali collective in Butler's note on slime molds, we would do well to remember that the Oankali, though far advanced in communicating across species lines and pushing beyond human notions of individuality and collectivity, were not without their coercive aspects. As "gene traders," the Oankali roamed the universe as scientific prospectors, mining for genetically valuable material. One of them, Jdahya, explains: "We do what you would call genetic engineering. . . . We *must* do it. . . . It is part of our reproduction, but it's much more deliberate than what any mated pair of humans have managed so far. . . . We're not hierarchical, you see. We never were. But we are powerfully acquisitive. We acquire new life—seek it, investigate it, manipulate it, sort it, use it" (Butler [1987] 1997, 39). The Oankali may claim to be nonhierarchical, but they approach the universe through frameworks of usability. As gene traders, they inhabit a capitalist, colonialist mindset of mergers and acquisitions in which "the merge" never quite takes place across even footing.

Butler's nuanced depiction of the Oankali as nonhierarchical but powerfully acquisitive is indicative of how her interest in the slime mold differs from that of entrepreneurial technoscience. Slime mold modeling in the service of capitalist technological innovation emphasizes efficiency, and its promise as projected by popular science media marvels at the alien intelligence of such a "primitive" species. The novelty of the story lies in the surprise humans have at nonhuman intelligence and how that intelligence can be harnessed to serve human interests. Such a relation reproduces a colonialist version of trans-species exchange and sustains fascination as a means of reinforcing human supremacy in species hierarchy.

At a moment when state and corporate project managers are looking to slime molds for direction in constructing self-organizing and cost-efficient networks in the real world, what can we learn differently from these problem-solving experimental subjects? Reading Butler's work through black, queer, decolonial studies provides a way to interrogate the processes of subjection into which slime molds have been called. There's a long history of scientific experimentation on people of color, and Butler's awareness of this racialized history leads her to a consideration of a trans-species set of solidarities. Lilith, the black protagonist of *Dawn*, understands this when she contemplates how the Oankali have subjected humans to a form of genetic experimentation:

> This was one more thing they had done to her body without her consent and supposedly for her own good. "We used to treat animals that way," she muttered bitterly. . . . "We did things to them—inoculations, surgery, isolation—all for their own good. We wanted them healthy and protected—sometimes so we could eat them later." (Butler [1987] 1997, 31)

Through Lilith's reflection on animal experimentation in the medical and meat industries, Butler asks us to consider what it means to rethink futurity from a multispecies undercommons. After all, Lilith likens Oankali gene trading not only to the meat industry but also to slave history: "Humans had done these things to captive breeders—all for a higher good, of course" (Butler [1987] 1997, 62). In slime mold, Butler may see a model for collective politics rather than merely problem-solving potentiality,[5] but she stops short of suggesting any sort of inherently liberatory ethos in collectivity. Though she takes interest in slime mold's plasmodial improprieties that confound hierarchical taxonomies, her characterization of the Oankali as "powerfully acquisitive" demonstrates the colonialist potentiality of collectivity, too. Perhaps Butler was also thinking of the 1958 film *The Blob*, which is to say communism,[6] though of course it's capitalism, too. We have witnessed how readily the World Bank has adapted the idea of the commons to suit global markets that actually serve private interests (Federici 2012).

In the Xenogenesis series, Butler's interest in the plasmodial improprieties of slime mold bump up against the matter of slavery—the rendering of human flesh as property. Reading Butler's *Dawn* as subaltern literature, Eva Cherniavsky invokes Hortense Spillers's theorization of the "theft of the body itself" to articulate the process by which "a body [is] rendered absolutely and impossibly improper insofar as it becomes (another's) property" (Cherniavsky 1996, 107). Oankali reproductive practice thoroughly sees dialectical relations of master/slave, self/other, and alien/human to their enmeshed ends. The Oankali, Cherniavsky continues, "practice reproduction as a form of corporate/corporeal impropriety, in which they perpetuate 'their' identity and agency by displacing themselves across the historical and territorial limits of Oankali culture" (1996, 108). In conversations about human and nonhuman ontologies, about intra-action and sympoiesis, black studies and decolonial theory offer much-needed reminders of how the category of the human even comes to be.

With this essay I mean to interject Butler's thinking beyond the human into a recent flurry of critical interest in Sylvia Wynter's interventions into Enlightenment humanism (Hantel 2015; Jackson 2013; McKittrick 2014). The category of the human, according to Wynter, catalyzed its liberation as a rights-bearing subject on the backs of slaves and many others relegated to the nonhuman. At a moment when the slime mold presents itself as a new material to think with, Butler's archive offers up another way to think beyond the human without flattening that concept into a universal given.

Conclusion: The Alien within the Human

I met my first slime mold not too long ago when it was time to put some mulch down in the northern woodlands of Vermont. I recoiled from its gelatinous movements, creeped out by its "dog vomit" masquerade and alien presentation. It may have been of this earth but it felt as though I were encountering an extraterrestrial, and I needed to unlearn the visceral disgust that came with this interspecies contact. Several months later, I made my first trip to the Octavia Butler papers, where I came across the note that launched this essay. The surprise I felt upon encountering the slime mold in the yard and the slime mold in the archive was quite similar. I have always understood the practice of reading science fiction as an exercise in thinking beyond the self. As a woman of color brought up in fairly conventional reading environments (at least in the classroom), I was asked constantly to understand from perspectives that were alien to me though they were often assumed to be universal.

This case study of the slime mold begins to reroute "the primacy of matter" in feminist theory through decolonial thought and queer-of-color critique (Coole and Frost 2010, 1).[7] If the turn to matter in philosophy asserts a realism beyond human ken, it engages a speculative realism that would have thinkers taking up slime mold as an object through which to imagine another ontology, beyond the human. Butler manages to do so without dissolving the human into the object—even as she wants to get to know it better. What she does is speculative fabulation, and I offer it up as a feminist queer science studies methodology.

Notes

1 I would like to thank the inspiring audience and participants at University of California San Diego's "Shaping Change: Remembering Octavia E. Butler" conference in June 2016.

2 At a moment when many in the humanities and social sciences are taking a turn to the nonhuman, I am not alone in looking to science fiction as a site of inquiry that has long been thinking beyond the human. Donna Haraway was the person who first articulated this connection in my own reading trajectory, but I also join Colin Milburn, Steven Shaviro, Rebekah Sheldon, McKenzie Wark, and several others in bringing together science fiction studies and conversations in the recent critical moves beyond the human.

3 Thanks to Sami Schalk, who brought this point to my attention during a June 4 Q&A session at the UCSD "Shaping Change" conference.

4 See also Tero et al. (2010), whose research on *Physarum polycephalum* led to the project featured in Sanders's *Wired* magazine article.

5 Indeed, the Oankali attribute the destruction of the human species to "two incompatible characteristics": Humans are intelligent, but we are also deeply hierarchical (Butler [1987] 1997, 37).

6 For a stunning account of the 1957 presidential prayer breakfast at which *The Blob* was conceived, see Jeff Sharlet's *The Family* (2008, 181).

7 Coole and Frost ask: "How could we ignore the power of matter and the ways it materializes in our ordinary experiences or fail to acknowledge the primacy of matter in our theories?"

Works Cited

Adamatzky, Andrew, and Genaro J. Martinez. 2013. "Bio-imitation of Mexican Migration Routes to the USA with Slime Mould on 3D Terrains." *Journal of Bionic Engineering* 10 (2): 242–250.

Ahmed, Sara. 2006. *Queer Phenomenology: Orientations, Objects, Others*. Durham: Duke University Press.

Bahng, Aimee. 2017. *Migrant Futures: Decolonizing Speculation in Financial Times*. Durham: Duke University Press.

Barad, Karen. 2007. *Meeting the Universe Halfway: Quantum Physics and the Entanglement of Matter and Meaning.* Durham: Duke University Press.
Bryant, Levi R., Nick Srnicek, and Graham Harman. 2011. "Towards a Speculative Philosophy." In *The Speculative Turn: Continental Materialism and Realism*, edited by Levi R. Bryant, Nick Srnicek, and Graham Harman, 1–18. Victoria: re.press.
Butler, Octavia E. 1988. "Notes on Organisms." December 31. Box 83, Folder 1625. Octavia E. Butler Papers. Huntington Library, San Marino, CA.
———. [1987] 1997. *Dawn.* New York: Warner Books.
———. 1993. *Parable of the Sower.* New York: Warner Books.
———. 2000. *Lilith's Brood.* New York: Aspect/Warner Books.
Cherniavsky, Eva. 1996. "Subaltern Studies in a US Frame." *boundary 2* 23 (2): 85–110.
Coole, Diana, and Samantha Frost. 2010. *New Materialisms: Ontology, Agency, and Politics.* Durham: Duke University Press.
Cooper, Melinda. 2011. *Life as Surplus: Biotechnology and Capitalism in the Neoliberal Era.* Seattle: University of Washington Press.
Federici, Silvia. 2012. "Feminism and the Politics of the Commons." In *The Wealth of the Commons: A World Beyond Market and State*, edited by David Bollier and Silke Helfrich. Amherst: Levellers Press.
Ferreira, Maria Aline. 2010. "Symbiotic Bodies and Evolutionary Tropes in the Work of Octavia Butler." *Science Fiction Studies* 37 (3): 401–415.
Gorby, Yuri. 2009. "Op-Ed: Microbes May Be More Networked than You Are." *Wired,* June 16. Accessed June 28, 2016. www.wired.com/2009/06/ftf-gorby/.
Hantel, Adam Maxlind. 2015. "Intergenerational Geographies of Race and Gender: Tracing the Confluence of Afro-Caribbean and Feminist Thought Beyond the Word of Man." PhD thesis, Rutgers. doi:10.7282/T3057HXS.
Haraway, Donna. 1991. *Simians, Cyborgs, and Women: The Reinvention of Nature.* New York: Routledge.
———. 2016. *Staying with the Trouble: Making Kin in the Chthulucene.* Durham: Duke University Press.
Hotz, Robert Lee. 2014. "Off the Wall: Scientists Get Prepped for Slime Mold Racing." *Wall Street Journal,* March 12. Accessed July 22, 2016.
Jackson, Zakiyyah Iman. 2013. "Animal: New Directions in the Theorization of Race and Posthumanism." *Feminist Studies* 39 (3): 669–685.
Keim, Brandon. 2008. "Complexity Theory in Icky Action: Meet the Slime Mold." *Wired,* February 15. Accessed June 28, 2016. www.wired.com/2008/02/complexity-th-1/.
Keller, Evelyn Fox. 1983. "The Force of the Pacemaker Concept in Theories of Aggregation in Cellular Slime Mold." *Perspectives in Biology and Medicine* 26 (4): 515–521.
McKittrick, Katherine. 2014. *Sylvia Wynter: On Being Human as Praxis.* Durham: Duke University Press.

Meillassoux, Quentin. 2008. *After Finitude: An Essay on the Necessity of Contingency*. New York: Continuum.
Milburn, Colin. 2010. "Modifiable Futures: Science Fiction at the Bench." *Isis* 101 (3): 560–569.
Puar, Jasbir. 2007. *Terrorist Assemblages: Homonationalism in Queer Times*. Durham: Duke University Press.
Sanders, Laura. 2010. "Slime Mold Grows Network Just Like Tokyo Rail System." *Wired*, January 22. Accessed June 28, 2016. www.wired.com/2010/01/slime-mold-grows-network-just-like-tokyo-rail-system/.
Sharlet, Jeff. 2008. *The Family: Power, Politics and Fundamentalism's Shadow Elite*. New York: HarperCollins.
Shaviro, Steven. 2016. *Discognition*. London: Repeater.
Subramaniam, Banu. 2014. *Ghost Stories for Darwin: The Science of Variation and the Politics of Diversity*. Urbana-Champaign: University of Illinois Press.
Tero, Atsushi, Seiji Takagi, Tetsu Saigusa, Kentaro Ito, Dan P. Bebber, Mark D. Fricker, Kenji Yumiki, Ryo Kobayashi, and Toshiyuki Nakagaki. 2010. "Rules for Biologically Inspired Adaptive Network Design." *Science* 327 (5964): 439–442.
Vint, Sherryl. 2010. "Animal Studies in the Era of Biopower." *Science Fiction Studies* 37 (3): 444–455.

Index

Feminist Technosciences

Rebecca Herzig and Banu Subramaniam, Series Editors

Figuring the Population Bomb: Gender and Demography in the Mid-Twentieth Century, by Carole R. McCann

Risky Bodies & Techno-Intimacy: Reflections on Sexuality, Media, Science, Finance, by Geeta Patel

Reinventing Hoodia: Peoples, Plants, and Patents in South Africa, by Laura A. Foster

Queer Feminist Science Studies: A Reader, edited by Cyd Cipolla, Kristina Gupta, David A. Rubin, and Angela Willey

Gender before Birth: Sex Selection in a Transnational Context, by Rajani Bhatia

www.ingramcontent.com/pod-product-compliance
Ingram Content Group UK Ltd.
Pitfield, Milton Keynes, MK11 3LW, UK
UKHW040803120225
454975UK00002B/64